# Boundary Elements and other Mesh Reduction Methods XXXII



WIT*PRESS*

WIT Press publishes leading books in Science and Technology.
Visit our website for the current list of titles.
www.witpress.com

WIT*eLibrary*

Home of the Transactions of the Wessex Institute.
Papers presented at BEM/MRM XXX II are archived in the WIT elibrary in volume 50 of WIT Transactions on Modelling and Simulation (ISSN 1743-355X).
The WIT electronic-library provides the international scientific community with immediate and permanent access to individual papers presented at WIT conferences.
http://library.witpress.com

# Thirty-Second International Conference on Boundary Elements and Other Mesh Reduction Methods

# BEM/MRM 2010

## Conference Chairman

**C.A. Brebbia**
*Wessex Institute of Technology, UK*

## International Scientific Advisory Committee




**Organised by**
*Wessex Institute of Technology, UK*

**Sponsored by**
*International Journal of Engineering Analysis with Boundary Elements (EABE)*

# Thirty-Second International Conference on Boundary Elements and Other Mesh Reduction Methods

# BEM/MRM 2010

### Conference Chairman

**C.A. Brebbia**
*Wessex Institute of Technology, UK*

### International Scientific Advisory Committee



**Organised by**
*Wessex Institute of Technology, UK*

**Sponsored by**
*International Journal of Engineering Analysis with Boundary Elements (EABE)*

# WIT Transactions

## Transactions Editor

**Carlos Brebbia**
Wessex Institute of Technology
Ashurst Lodge, Ashurst
Southampton SO40 7AA, UK
Email: carlos@wessex.ac.uk

## Editorial Board

# Boundary Elements and other Mesh Reduction Methods XXXII

**Editor**

**C.A. Brebbia**
*Wessex Institute of Technology, UK*

WIT*PRESS* **Southampton, Boston**

**Editor:**

**C.A. Brebbia**
*Wessex Institute of Technology, UK*

Published by

**WIT Press**
Ashurst Lodge, Ashurst, Southampton, SO40 7AA, UK
Tel: 44 (0) 238 029 3223; Fax: 44 (0) 238 029 2853
E-Mail: witpress@witpress.com
http://www.witpress.com

For USA, Canada and Mexico

**Computational Mechanics Inc**
25 Bridge Street, Billerica, MA 01821, USA
Tel: 978 667 5841; Fax: 978 667 7582
E-Mail: infousa@witpress.com
http://www.witpress.com

British Library Cataloguing-in-Publication Data

A Catalogue record for this book is available
from the British Library

ISBN: 978-1-84564-470-3
ISSN: (print) 1746-4064
ISSN: (on-line) 1743-355X

*The texts of the papers in this volume were set individually by the authors or under their supervision. Only minor corrections to the text may have been carried out by the publisher.*



Printed in Great Britain by Martins the Printers.

# Preface

Looking back to 1978 when the first of these conferences took place, it is rewarding to see how much the Boundary Element Method has developed in the last three decades (or – to be precise – 33 years if one dates its beginning from when the first paper on BEM was published*).

What is also remarkable is that the Method originated a whole series of "derivatives" – if I can use that word – dealing with mesh reduction techniques. It was indeed the quest for mesh reduction that was the motivation for the development of the BEM. The fact that the Method was based on boundary integral equations was driven by engineering needs rather than by their mathematical beauty and elegance.

Given the continuous emergence of new techniques, it is important nowadays to bear those concepts in mind, focusing on the user's needs rather than the interests of the developers. For any new mesh reduction method to be truly successful, it must perform well in engineering terms. Analysis techniques which are not reliable and robust are bound to fail, regardless of their higher accuracy and excellent performance in an academic environment.

Another important quality that any new computational technique needs to offer to industrial users is ease of application. Nowadays, this involves good interfacing with existing modelling systems as well as the need to be independent of the setting up of parameters or coefficients which, for all their mathematical importance, are meaningless to the user.

It is now a good time to start looking at the new generation of analytical tools based on the substantial amount of work carried out in the last few years. New ideas seem to develop within a "generation" period. This seems to be the case if

*Brebbia, C.A. and Dominguez, J. "The boundary element method for potential problems" *Appl. Math. Modelling*, **1**, 372–8 (1977).

we look back at the first papers on FEM (1955) and BEM (1977). The Proceedings of this series of Conferences, which encourages the presentation of novel ideas, may hold the key to the future of engineering analysis, particularly the volumes published in the last 10 years or so.

We leave to the new generation the task of unravelling the future.

Carlos A. Brebbia
The New Forest 2010

# Contents

**Section 3: Dynamics and vibrations**



**Section 4: Electrical engineering and electromagnetics**



**Section 5: Advanced meshless and mesh reduction methods**

**Section 6: Advanced formulations**



**Section 7: Computational techniques**

# Section 1
# Fluid flow

# Simulation of flow of nanofluids by BEM

J. Ravnik & L. Škerget
*University of Maribor, Faculty of Mechanical Engineering, Slovenia*

## Abstract

Heat transfer is one of the major engineering problems, which has to be addressed when progress is made in engineering technology. Low thermal conductivity of fluids such as water or oil, led to the introduction of nanofluids – stable suspensions of nanosized particles in base fluid. Nanofluids have very high thermal conductivities at very low nanoparticle concentrations and are able to substantially increase heat transfer.

Numerical solver for simulation of flow and heat transfer of nanofluids was developed using Boundary Element Method. Velocity-vorticity formulation of Navier-Stokes equations is proposed for nanofluids and solved by a combination of single domain and sub-domain boundary element method. The developed solver was used to simulate three-dimensional natural convection of nanofluids. Results of simulations show that a substantial heat transfer enhancement can be observed and attributed to the usage of nanofluids. We observed that the nanofluid particle volume fraction also plays in important role  increasing the volume fraction increases the heat transfer for all Rayleigh number values considered.
*Keywords: nanofluids, boundary element method, natural convection.*

## 1 Introduction

Nanofluid is a suspension consisting of uniformly dispersed and suspended nanometre-sized (10–50 nm) particles in base fluid, pioneered by Choi [1]. Nanofluids are used to replace low thermal conductivity working fluids such as water, oil or ethylene glycol. Nanofluids have very high thermal conductivities at very low nanoparticle concentrations.

Buoyancy induced flow and heat transfer is an important phenomenon used in various engineering systems. Increased thermal conductivity of nanofluids plays a crucial role in the enhancement of heat transfer especially in slow, low Rayleigh

WIT Transactions on Modelling and Simulation, Vol 50, © 2010 WIT Press
www.witpress.com, ISSN 1743-355X (on-line)
doi:10.2495/BE100011

number, flows. In this regime, usage of nanofluids instead of classical working fluids, is most advantageous.

Numerous numerical methods have been proposed to simulate the natural convection phenomena of pure fluids. Our, boundary element based methods, was presented in Ravnik et al. [2]. We employ the velocity-vorticity formulation of Navier-Stokes equations coupled with the energy equation. The unknown field functions are the velocity, vorticity and temperature. We use single domain BEM to solve the kinematics equation for boundary vorticity values (Škerget et al. [3]). Apart for the boundary vorticity values, all other flow fields are solved for by subdomain BEM (Popov et al. [4]). Subdomain BEM solution of a partial differential equation leads to an over-determined sparse system of linear equations. A sparse system enables fast algebraic operations and does not require a lot of storage.

In this work we implemented the velocity vorticity formulation written for nanofluids and derived a BEM based numerical algorithm to find a solution.

## 2 Governing equations

We consider water based nanofluids. Their thermophysical properties are given in Table 1. Effective properties of the nanofluid are: density $\rho_{nf}$, dynamic viscosity $\mu_{nf}$, heat capacitance $(c_p)_{nf}$, thermal expansion coefficient $\beta_{nf}$ and thermal conductivity $k_{nf}$, where subscript *nf* is used to denote effective i.e. nanofluid properties. The properties are all assumed constant throughout the flow domain. The non-dimensional velocity-vorticity formulation of Navier-Stokes equations for simulation of nanofluids consists of the kinematics equation, the vorticity transport equation and the energy equation:

$$\nabla^2\vec{v} + \vec{\nabla} \times \vec{\omega} = 0, \tag{1}$$

$$\frac{\partial \vec{\omega}}{\partial t} + (\vec{v}\cdot\vec{\nabla})\vec{\omega} = (\vec{\omega}\cdot\vec{\nabla})\vec{v} + Pr\frac{\mu_{nf}}{\mu_f}\frac{\rho_f}{\rho_{nf}}\nabla^2\vec{\omega} - PrRa\frac{\beta_{nf}}{\beta_f}\vec{\nabla}\times T\vec{g}, \tag{2}$$

Table 1: Thermophysical properties of water based nanofluids (Oztop and Abu-Nada [6]).

| | pure water | $Cu$ | $Al_2O_3$ | $TiO_2$ |
|---|---|---|---|---|
| $c_p[J/kgK]$ | 4179 | 385 | 765 | 686.2 |
| $\rho[kg/m^3]$ | 997.1 | 8933 | 3970 | 4250 |
| $k[W/mK]$ | 0.613 | 400 | 40 | 8.9538 |
| $\beta[\cdot 10^{-5}K^{-1}]$ | 21 | 1.67 | 0.85 | 0.9 |
| $\alpha[\cdot 10^{-7}m^2/s]$ | 1.47 | 1163 | 131.7 | 30.7 |

$$\frac{\partial T}{\partial t} + (\vec{v} \cdot \vec{\nabla})T = \frac{k_{nf}}{k_f} \frac{(\rho c_p)_f}{(\rho c_p)_{nf}} \nabla^2 T. \tag{3}$$

The flow and heat transfer of a nanofluid is defined by specifying the pure fluid Rayleigh $Ra$ and Prandtl $Pr$ number values. The nanofluid properties are evaluated using the following models (Khanafer et al. [5]): density from (4), viscosity from (5), thermal capacity from (6), thermal expansion from (7) and thermal conductivity from (8):

$$\rho_{nf} = (1 - \varphi)\rho_f + \varphi \rho_s, \tag{4}$$

$$\mu_{nf} = \frac{\mu_f}{(1 - \varphi)^{2.5}}, \tag{5}$$

$$(\rho c_p)_{nf} = (1 - \varphi)(\rho c_p)_f + \varphi(\rho c_p)_s, \tag{6}$$

$$\beta_{nf} = \beta_f \left[ \frac{1}{1 + \frac{(1-\varphi)\rho_f}{\varphi \rho_s}} \frac{\beta_s}{\beta_f} + \frac{1}{1 + \frac{\varphi}{1-\varphi} \frac{\rho_s}{\rho_f}} \right], \tag{7}$$

$$k_{nf} = k_f \frac{k_s + 2k_f - 2\varphi(k_f - k_s)}{k_s + 2k_f + \varphi(k_f - k_s)}. \tag{8}$$

## 3 Numerical method

The algorithm used to solve the set of governing equations (1)–(3) is devised as follows. Either Dirichlet or Neumann type boundary conditions for velocity and temperature must be known. In this paper we use the no-slip boundary condition on all solid walls and prescribe temperature or temperature flux. Boundary conditions for vorticity are unknown and are calculated as a part of the algorithm. The following steps are performed.

1. Use models (4)–(8) to calculate ratios of nanofluid to pure fluid material properties
2. Calculate vorticity values on the boundary by single domain BEM from the kinematics equation (1).
3. Calculate velocity values by sub-domain BEM from the kinematics equation (1).
4. Calculate temperature values by sub-domain BEM from the energy equation (3).
5. Calculate vorticity values in the domain by sub-domain BEM from the vorticity transport equation (2).
6. Check convergence. If all flow fields converged to $10^{-6}$ stop, else go to 2.

The three-dimensional solver capable of simulating flow and heat transfer by solving velocity-vorticity formulation of Navier-Stokes equations by a combination of single and sub-domain BEM was developed by Ravnik et al. [2, 7].

The solver has been adapted for simulation of flow and heat transfer of nanofluids. Governing equations are written in integral form. The kinematics equation is

$$c(\vec{\xi})\vec{n}(\vec{\xi}) \times \vec{v}(\vec{\xi}) + \vec{n}(\vec{\xi}) \times \int_{\Gamma} \vec{v}\vec{\nabla}u^{\star} \cdot \vec{n}d\Gamma$$

$$= \vec{n}(\vec{\xi}) \times \int_{\Gamma} \vec{v} \times (\vec{n} \times \vec{\nabla})u^{\star}d\Gamma + \vec{n}(\vec{\xi}) \times \int_{\Omega} (\vec{\omega} \times \vec{\nabla}u^{\star})d\Omega, \qquad (9)$$

where $\vec{\xi}$ is the source or collocation point, $\vec{n}$ is a vector normal to the boundary, pointing out of the domain and $u^{\star}$ is the fundamental solution for the diffusion operator: $u^{\star} = 1/4\pi|\vec{\xi} - \vec{r}|$. $c(\vec{\xi})$ is the geometric factor defined as $c(\vec{\xi}) = \alpha/4\pi$, where $\alpha$ is the inner angle with origin in $\vec{\xi}$. In order to write a linear system of equations for the unknown boundary vorticity values, we set the source point into every boundary node of the whole computational domain. This yields a full system matrix where number of rows and columns is equal to number of boundary nodes. It is solved using a LU decomposition method.

The partial derivative with respect to time in the kinetics equations is approximated by second order three point finite difference scheme $\partial T/\partial t = (3T - 4T' + T'')/2\Delta t$, where $\Delta t$ is the time step and prime and double prime denote field functions in two earlier time steps. The final forms of vorticity transport and energy equation are

$$c(\vec{\xi})\omega_j(\vec{\xi}) + \int_{\Gamma} \omega_j \vec{\nabla}u^* \cdot \vec{n}d\Gamma = \int_{\Gamma} u^* q_j d\Gamma + \frac{1}{Pr}\frac{\mu_f}{\mu_{nf}}\frac{\rho_{nf}}{\rho_f} \cdot$$

$$\cdot \left( \int_{\Gamma} \vec{n} \cdot \{u^*(\vec{v}\omega_j - \vec{\omega}v_j)\} d\Gamma - \int_{\Omega} (\vec{v}\omega_j - \vec{\omega}v_j) \cdot \vec{\nabla}u^* d\Omega \right)$$

$$-Ra\frac{\beta_{nf}}{\beta_f}\frac{\mu_f}{\mu_{nf}}\frac{\rho_{nf}}{\rho_f} \left( \int_{\Gamma} (u^{\star}T\vec{g} \times \vec{n})_j d\Gamma + \int_{\Omega} (T\vec{\nabla} \times u^{\star}\vec{g})_j d\Omega \right)$$

$$+\frac{1}{Pr}\frac{\mu_f}{\mu_{nf}}\frac{\rho_{nf}}{\rho_f}\frac{1}{2\Delta t}\int_{\Omega} (3\omega_j - 4\omega_j' + \omega_j'')u^* d\Omega, \qquad (10)$$

$$c(\vec{\xi})T(\vec{\xi}) + \int_{\Gamma} T\vec{\nabla}u^* \cdot \vec{n}d\Gamma = \int_{\Gamma} u^* T_q d\Gamma$$

$$+\frac{k_f}{k_{nf}}\frac{(\rho c_p)_{nf}}{(\rho c_p)_f} \left( \int_{\Gamma} \vec{n} \cdot \{u^*(\vec{v}T)\} d\Gamma - \int_{\Omega} (\vec{v}T) \cdot \vec{\nabla}u^* d\Omega \right)$$

$$+\frac{k_f}{k_{nf}}\frac{(\rho c_p)_{nf}}{(\rho c_p)_f}\frac{1}{2\Delta t}\int_{\Omega} (3T - 4T' + T'')u^* d\Omega, \qquad (11)$$

where $\omega_j$ is the $j^{th}$ component of vorticity.

In the subdomain BEM method we make a mesh of the entire domain $\Omega$ and name each mesh element a subdomain. Equation (10) is written for each of the subdomains. In order to obtain a discrete version of (10) we use shape functions to interpolate field functions and flux across the boundary and inside of the subdomain. In this work we used hexahedral subdomains, which enable continuous quadratic interpolation of field functions. On each boundary element we interpolate the flux using discontinuous linear interpolation scheme. By using discontinuous interpolation we avoid flux definition problems in corners and edges. A function, e.g. temperature, is interpolated over a boundary elements as $T = \sum \varphi_i T_i$, inside each subdomain as $T = \sum \Phi_i T_i$, while flux is interpolated over boundary elements as $q = \sum \phi_i q_i$. The following integrals must be calculated:

$$[H] = \int_\Gamma \varphi_i \vec{\nabla} u^\star \cdot \vec{n} d\Gamma, \quad [G] = \int_\Gamma \phi_i u^\star d\Gamma, \quad [\vec{A}] = \int_\Gamma \varphi_i \vec{n} u^\star d\Gamma, \tag{12}$$

$$[B] = \int_\Omega \Phi_i u^\star d\Omega, \qquad [\vec{D}] = \int_\Omega \Phi_i \vec{\nabla} u^\star d\Omega. \tag{13}$$

The square brackets denote integral matrices. In order to calculate the integrals, a Gaussian quadrature algorithm is used. Calculation of the free coefficient $c(\vec{\xi})$ is preformed indirectly considering rigid body movement problem solution. The calculated $c(\vec{\xi})$ are added to the diagonal terms of the $[H]$ matrix.

The source point is set to all function and flux node in each subdomain. By letting curly brackets denote vectors of nodal values of field functions, we may write the discrete vorticity transport equation for $x$ component as:

$$\begin{aligned}
[H]\{\omega_x\} &= [G]\{q_x\} \\
&+ \frac{1}{Pr}\frac{\mu_f}{\mu_{nf}}\frac{\rho_{nf}}{\rho_f}\left([A_y]\{v_y\omega_x - \omega_y v_x\} + [A_z]\{v_z\omega_x - \omega_z v_x\}\right) \\
&- \frac{1}{Pr}\frac{\mu_f}{\mu_{nf}}\frac{\rho_{nf}}{\rho_f}\left([D_y]\{v_y\omega_x - \omega_y v_x\} + [D_z]\{v_z\omega_x - \omega_z v_x\}\right) \\
&+ Ra\frac{\beta_{nf}}{\beta_f}\frac{\mu_f}{\mu_{nf}}\frac{\rho_{nf}}{\rho_f}\left(g_z[A_y]\{T\} - g_y[A_z]\{T\} - g_z[D_y]\{T\} + g_y[D_z]\{T\}\right) \\
&+ \frac{1}{Pr}\frac{\mu_f}{\mu_{nf}}\frac{\rho_{nf}}{\rho_f}\frac{1}{2\Delta t}[B](3\{\omega_x\} - 4\{\omega_x'\} + \{\omega_x''\}).
\end{aligned} \tag{14}$$

Equations for $y$ and $z$ components are obtained analogously. Similarly, the discrete counterpart of the energy equation (11) is:

$$\begin{aligned}
[H]\{T\} &= [G]\{q_T\} \\
&+ \frac{k_f}{k_{nf}}\frac{(\rho c_p)_{nf}}{(\rho c_p)_f}\left([A_x]\{v_x T\} + [A_y]\{v_y T\} + [A_z]\{v_z T\}\right) \\
&- \frac{k_f}{k_{nf}}\frac{(\rho c_p)_{nf}}{(\rho c_p)_f}\left([D_x]\{v_x T\} + [D_y]\{v_y T\} + [D_z]\{v_z T\}\right) \\
&+ \frac{k_f}{k_{nf}}\frac{(\rho c_p)_{nf}}{(\rho c_p)_f}\frac{1}{2\Delta t}[B](3\{T\} - 4\{T'\} + \{T''\}),
\end{aligned} \tag{15}$$

where $\{q_T\}$ is a nodal vector of temperature flux.

 
www.witpress.com, ISSN 1743-355X (on-line)

Since neighbouring subdomains share nodes, the resulting systems of linear equations are over-determined. After taking into account the boundary conditions, we solve them using a least squares solver (Paige and Saunders [8]). All integrals depend only on the shape of subdomains and as such may be calculated only once, prior to the start of the nonlinear iterative process.

### 3.1 Acceleration of computation

The numerical algorithm solves the above equations sequentially within a time step. At the end of each iteration, RMS difference between flow fields of current and previous iteration is calculated. When RMS difference for all flow fields reaches $\epsilon_{err} = 10^{-6}$, we stop to iterate. Within each iteration 7 over-determined systems of linear equations must be solved (three for domain velocity, three for domain vorticity and one for temperature). The LSQR solver [8] with diagonal preconditioning is used. The solver requires a large number of iterations in order to converge to a predefined convergence criteria $\epsilon$. The original algorithm, as proposed by Ravnik et al. [7], uses a constant convergence criteria. Value, which is 10 times less than the required RMS criteria $\epsilon_{err}$ was usually used, i.e. $\epsilon = \epsilon_{err}/10 = 10^{-7}$.

We argue, that it is not necessary to keep $\epsilon = 10^{-7}$ during the whole iterative process, since at the beginning, when RMS differences $\epsilon_{err}$ are large, we do not require a very accurate solution of linear systems of equations. Considering this $\epsilon$ may set larger. Since the number of iterations of the solver of linear systems of equation depends strongly on the required accuracy and since the most CPU time is used in these routines, we anticipate a large decrease of CPU time.

In order to accelerate the solver, we used the following steps. Instead of keeping $\epsilon$ constant, we change its value every iteration. We set the range for $\epsilon$ between $\epsilon_{min} = 10^{-7}$ and $\epsilon_{max} = 10^{-3}$. We introduce a parameter $R$; $1 \leq R \leq 100$. At the end of each iteration, we calculate RMS differences $\epsilon_{err}$ for all flow fields and use the following algorithm to determine the new $\epsilon$:

```
DO ∀ equations
   ε_old = ε
   IF (ε_err/R.LT.ε) THEN ε=ε_err/R ELSE ε=ε_max
   IF (ε_old.LT.ε) ε = ε_old
   IF (ε.LT.ε_min) ε = ε_min
END DO
```

We keep the linear solver accuracy between the minimum and maximum value at $\epsilon = \epsilon_{err}/R$. In the algorithm we do not allow for decrease of linear solver accuracy. If at some stage during the nonlinear loop RMS difference increases, the linear solver accuracy is not increased.

In order to test the algorithm, natural convection of air was simulated. Simulations were run for $Ra = 10^3, 10^4$ and $10^5$ using parameter $R$ ranging between $R = 0.1$ and $R = 100$. Table 2 shows the number of iterations of the LSQR solver [8], CPU time and the number of iterations of the nonlinear loop.

Table 2: Comparison of convergence performance and the number of iterations of LSQR solver [8] for different values of parameter $R$. All simulations were run until RMS difference reached $\epsilon_{err} = 10^{-6}$.

| $\epsilon_{min}$ | $\epsilon_{max}$ | $R$ | LSQR nit [$\cdot 10^3$] | CPU time [min] | No. non-linear iterations |
|---|---|---|---|---|---|
| $Ra = 10^3$ | | | | | |
| $10^{-7}$ | $10^{-7}$ | - | 103.5 | 75 | 131 |
| $10^{-7}$ | $10^{-3}$ | 100 | 73.6 | 55 | 132 |
| $10^{-7}$ | $10^{-3}$ | 10 | 57.2 | 45.7 | 138 |
| $10^{-7}$ | $10^{-3}$ | 1 | 37.9 | 35.5 | 154 |
| $Ra = 10^4$ | | | | | |
| $10^{-7}$ | $10^{-7}$ | - | 148.8 | 108 | 205 |
| $10^{-7}$ | $10^{-3}$ | 10 | 79.9 | 66.3 | 207 |
| $Ra = 10^5$ | | | | | |
| $10^{-7}$ | $10^{-7}$ | - | 759.9 | 570 | 1235 |
| $10^{-7}$ | $10^{-3}$ | 10 | 410.6 | 363 | 1255 |
| $10^{-7}$ | $10^{-3}$ | 1 | 202.9 | 228 | 1179 |
| $10^{-7}$ | $10^{-3}$ | 0.1 | 150.6 | 241 | 1774 |

Within each non-linear loop iteration, seven linear systems of equations must be solved using LSQR solver. The table presents cumulative values for number of LSQR iterations.

At $Ra = 10^3$, the total number of iterations of the LSQR solver required for the computation to converge at constant $\epsilon$ drops to more than one half of its value when using newly proposed dynamic solver accuracy algorithm with $R = 1$. At the same time, the CPU time of the whole non-linear loop is also decreased by one half. At higher Rayleigh number values even better decrease of CPU time is observed. The gain in CPU time seems to be unaffected by the nonlinearity of the problem, i.e. the Rayleigh number. The number of non-linear iterations needed to reach the solution is also stable. Based on this analysis we decided to use dynamic solver accuracy algorithm and we chose the value $R = 10$ for all further analyses.

## 4 Test case

We used the developed numerical algorithm to simulate natural convection of nanofluids in a cubic cavity. The cavity is filled with fluid and subjected to a temperature difference on two opposite vertical sides, while the other four are adiabatic. Water ($Pr = 6.2$) and water based nanofluids (Table 1) for several

Rayleigh number values were considered. In our previous work (Ravnik et al. [2]) we examined this problem using air as the working fluid. We established, by comparison with other authors (Tric et al. [9]), that a grid with $25^3$ nodes was sufficient. In this paper we increased the grid density to $41^3$ nodes in order to further increase the accuracy of computations. Nodes were concentrated towards the hot and cold walls in such a way that the ratio between the largest and the smallest element length was 7.

In order to compare effectiveness of nanofluids in different simulations, wall heat flux is calculated. Usually, the heat flux $\dot{Q}$ is expressed in terms of pure fluid thermal conductivity, characteristic flow scales and a non-dimensional Nusselt number, i.e. $\dot{Q} = k_f L \Delta T \cdot Nu$. The Nusselt number, $Nu$, is defined as the integral of the temperature flux through a wall. For a nanofluid, it is written as

$$Nu = \frac{k_{nf}}{k_f} \int_\Gamma \vec{\nabla} T \cdot \vec{n} dA, \tag{16}$$

where $\Gamma$ is the surface through which we calculate the heat flux and $\vec{n}$ is a unit normal to this surface. We study local variation of heat flux using the local Nusselt number defined as $Nu_l(x, y, z) = k_{nf}/k_f \vec{\nabla} T \cdot \vec{n}$.

### 4.1 Results

Applying a temperature difference on two opposite walls of an otherwise insulated cavity starts up natural convection producing a large vortex in the main part of the enclosure. At low Rayleigh number values the vortex is weak and the heat is transferred predominately by diffusion. Convection dominates at $Ra = 10^6$, where temperature stratification may be observed. The flow becomes unsteady for higher $Ra$ values with vortices forming along the hot and cold walls. Due to high thermal conductivity of nanofluids we expect to observe the largest improvement in heat transfer of nanofluids for cases where diffusion play a non negligible role. Thus, simulations were performed for Rayleigh number values between $Ra = 10^3$ and $Ra = 10^6$ for three nanofluids. Two solid nanoparticle volume fractions in nanofluids were considered: $\varphi = 0.1$ and $\varphi = 0.2$.

Nusselt number values for the natural convection in a cube are shown in Table 3. Using water based nanofluids instead of pure water increases heat transfer in all cases. For low Rayleigh number, where diffusion is the predominant heat transfer mechanism, the enhancement is the largest. For $Cu$ nanofluid at $Ra = 10^3$ we observe an 27.2% increase in heat transfer for $\varphi = 0.1$ and 64.1% for $\varphi = 0.2$. Similar findings were reported by Abu-Nada and Oztop [10] for 2D inclined cavity case.

As the Rayleigh number increases, convection becomes the dominant heat transfer mechanism, while diffusion is negligible. Thus, the increased thermal conductivity of nanofluids plays a less important role in the overall heat balance. All nanofluids exhibit smaller heat transfer enhancement as compared to the low Rayleigh number case. At $Ra = 10^6$ $Cu$ nanofluid increases heat transfer at $\varphi = 0.1$ for 11.6% and at $\varphi = 0.2$ for 21.6%.

Table 3: Nusselt number values for the natural convection in a cube. Solid particle volume fraction is denoted by $\varphi$.

| $Ra$ | Water | Water+$Cu$ | | Water+$Al_2O_3$ | | Water+$TiO_2$ | |
|---|---|---|---|---|---|---|---|
| | pure | $\varphi = 0.1$ | $\varphi = 0.2$ | $\varphi = 0.1$ | $\varphi = 0.2$ | $\varphi = 0.1$ | $\varphi = 0.2$ |
| $10^3$ | 1.071 | 1.363 | 1.758 | 1.345 | 1.718 | 1.297 | 1.598 |
| $10^4$ | 2.078 | 2.237 | 2.381 | 2.168 | 2.244 | 2.115 | 2.132 |
| $10^5$ | 4.510 | 4.946 | 5.278 | 4.806 | 4.968 | 4.684 | 4.732 |
| $10^6$ | 9.032 | 10.08 | 10.98 | 9.817 | 10.39 | 9.556 | 9.871 |

In our geometry the hot and cold walls face each other in $x$ direction. Thus, the main vortex, which is induced by the onset of natural convection is located in the $y$ plane. We chose the $y = 0.5H$ plane to study the two velocity profiles: vertical velocity $v_z(x, 0.5H, 0.5H)$ and horizontal velocity $v_x(0.5H, 0.5H, z)$ across the centre of the enclosure. The comparison of profiles for $Ra = 10^3$ and $Ra = 10^6$ for water and nanofluids is shown in Figure 1.

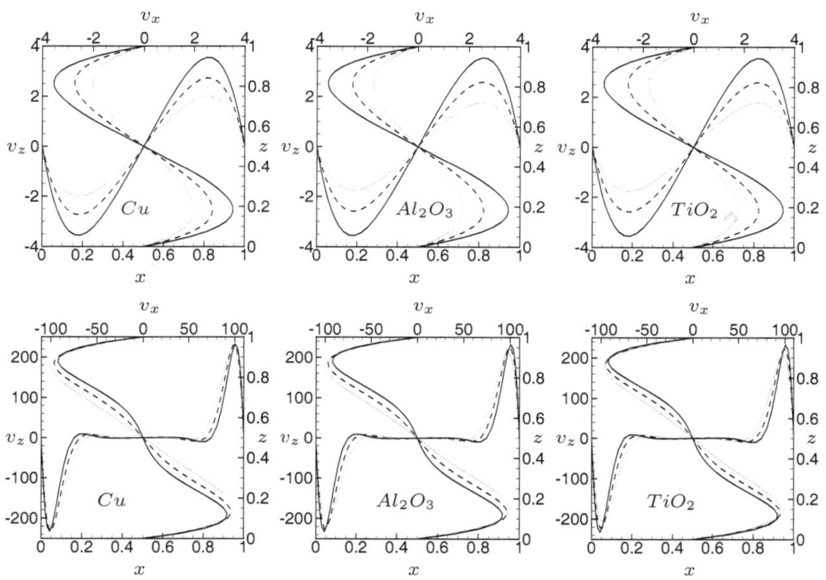


Figure 1: Velocity profiles $v_x(z)$ and $v_z(x)$ through a centre of the $y = 0.5H$ plane for natural convection in a differentially heated cubic cavity. Top row $Ra = 10^3$, bottom row $Ra = 10^6$. Solid line denotes pure water, dashed line $\varphi = 0.1$ nanofluid and dotted line $\varphi = 0.2$ nanofluid.

When diffusion dominates ($Ra = 10^3$) we observe that pure water reaches the highest velocities, while addition of solid particles slows down the flow. The decreased velocity results in decreased convective heat transfer. However, since in this regime the majority of heat is transported by diffusion, the decrease due to lower velocity is almost negligible and the overall heat transfer of nanofluids is still very large due to higher thermal conductivity of a nanofluid. In the $Ra = 10^6$ case convection dominates. Here we observe that velocities reached by nanofluids are higher that the velocities of pure water. Thus, using nanofluids, the velocity profiles, and in consequence, temperature profiles and heat transfer are increased. The relative increase of heat transfer in the convection dominated case is smaller that the increase in the diffusion dominated case because the increased thermal conductivity does not play an important role in the convection dominated heat transfer. When comparing velocity profiles between different nanofluids we observe only slight differences. The $Cu$ nanofluid reaches the highest velocities, while the highest velocities for $Al_2O_3$ nanofluid are about $4\%$ lower and for $TiO_2$ approximately $9\%$ lower.

Figure 2 displays temperature contours in the central $y = 0.5H$ plane. Comparing the temperature fields for different nanofluids we observe almost identical temperature distribution in the central part of the enclosure. Differences are larger closer to the walls, although their magnitude is still small. Since heat transfer depends on the temperature gradient on the walls, we examined the temperature contour closest to the hot and cold walls.

Heat flux distribution on a profile across the hot wall is shown in Figure 3. The heat flux is smaller at the top of the hot wall and larger at the bottom, since the hot fluid is rising and decreasing the temperature gradient at the top. Comparing the $Ra = 10^3$ and $Ra = 10^6$ cases, we see that the heat transfer enhancement is, relatively speaking, larger in the diffusion dominated $Ra = 10^3$ case than the convection dominated $Ra = 10^6$ case.

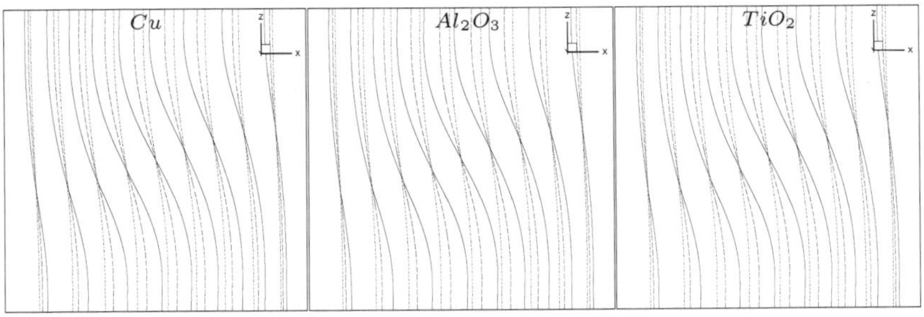


Figure 2: Temperature contours on the central $y = 0.5H$ plane. Contour values are -0.4(0.1)0.4; $Ra = 10^3$. Solid line denotes pure water, dashed line $\varphi = 0.1$ nanofluid and dotted line $\varphi = 0.2$ nanofluid.

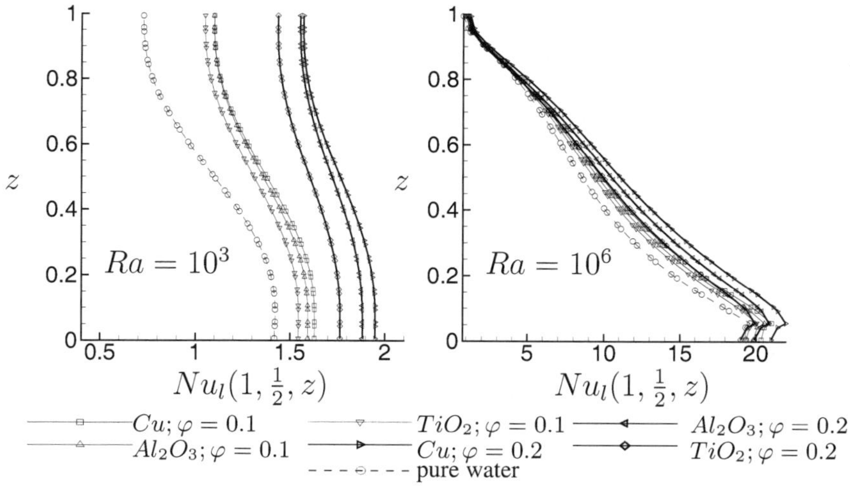


Figure 3: Heat flux shown in terms of local Nusselt number on a profile across hot wall on the $y = 0.5H$ plane.

## 5 Summary

The paper presents a numerical method for the simulation of flow of nanofluids. The method is based on the BEM solution of the velocity vorticity formulation of Navier-Stokes equations. A combination of single domain and sub-domain BEM was used. A dynamic solver accuracy algorithm was implemented to speed up simulations.

The developed algorithm was used to simulate natural convection in a three-dimensional differentially heated cavity of three types of water based nanofluids. Heat transfer, expressed with Nusselt number values, was presented for water and three nanofluids ($Cu$, $Al_2O_3$ and $TiO_2$). Results show that using water based nanofluids instead of pure water enhances heat transfer and that the enhancement is largest when diffusion is the dominant heat transfer mechanism. Simulations also revealed that heat transfer enhancement grows with increasing solid nano-particle volume fraction in the nanofluid.

## References


[1] Choi, S.U.S., Enhancing thermal conductivity of fluids with nanoparticles. *Develop Appl Non Newtonian Flows*, **66**, pp. 99–106, 1995.

[2] Ravnik, J., Škerget, L. & Žunič, Z., Velocity-vorticity formulation for 3D natural convection in an inclined enclosure by BEM. *Int J Heat Mass Transfer*, **51**, pp. 4517–4527, 2008.

[3] Škerget, L., Hriberšek, M. & Žunič, Z., Natural convection flows in complex

cavities by BEM. *Int J Num Meth Heat & Fluid Fl*, **13**, pp. 720–735, 2003.

[4] Popov, V., Power, H. & Škerget, L., (eds.) *Domain Decomposition Techniques for Boundary Elements: Applications to fluid flow*. WIT press, 2007.

[5] Khanafer, K., Vafai, K. & Lightstone, M., Buoyancy-driven heat transfer enhancement in a two-dimensional enclosure utilizing nanofluids. *Int J Heat Mass Transfer*, **46**, pp. 3639–3653, 2003.

[6] Oztop, H.F. & Abu-Nada, E., Natural convection of water-based nanofluids in an inclined enclosure with a heat source. *Int J Heat Fluid Flow*, **29**, pp. 1326–1336, 2008.

[7] Ravnik, J., Škerget, L. & Žunič, Z., Combined single domain and subdomain BEM for 3D laminar viscous flow. *Eng Anal Bound Elem*, **33**, pp. 420–424, 2009.

[8] Paige, C.C. & Saunders, M.A., LSQR: An algorithm for sparse linear equations and sparse least squares. *ACM Transactions on Mathematical Software*, **8**, pp. 43–71, 1982.

[9] Tric, E., Labrosse, G. & Betrouni, M., A first incursion into the 3D structure of natural convection of air in a differentially heated cubic cavity, from accurate numerical simulations. *Int J Heat Mass Transfer*, **43**, pp. 4034–4056, 2000.

[10] Abu-Nada, E. & Oztop, H.F., Effects of inclination angle on natural convection in enclosures filled with cuwater nanofluid. *Int J Heat Fluid Fl*, **30**, pp. 669–678, 2009.

# Boundary element solution of thermal creeping flow in a nano single mixer

C. Nieto[1], H. Power[2] & M. Giraldo[1]
[1]*Universidad Pontificia Bolivariana, Grupo de Energía y Termodinámica, Facultad de Ingeniería Mecánica, Colombia*
[2]*Faculty of Engineering, Department of Mechanical, Materials and Manufacturing Engineering, The University of Nottingham, UK*

## Abstract

In order to employ continuum models in the analysis of the flow behaviour of a viscous Newtonian fluid in micro scale devices, it is necessary to consider at the wall surfaces appropriate slip boundary conditions instead of the classical non-slip condition. The slip behaviour in the case of micro fluid flow of rarefied gases is associated with the combined effect of reduction in momentum transfer due to the reduction in the number of molecules (shear creep) and the thermal creep or transpiration, which as a consequence of inequalities in temperatures, forces the fluid to slide over a surface from colder to hotter regions. In this work a boundary integral equation formulation for Stokes slip flow, based on the normal and tangential projection of the Green's integral representational formulae for the Stokes velocity field, which directly incorporates into the integral equations the local tangential shear rate and heat flux at the wall surfaces, is presented. The tangential heat flux is evaluated in terms of the tangential gradient of the temperature integral representational formulae presenting singularity of the Cauchy type, which are removed by the use of an auxiliary field. These formulations are used to simulate a Single rotor mixer and analyze the combined effect of both shear and thermal creep effects over mixer performance.

*Keywords: linear slip boundary conditions, boundary element method, rotating mixers, thermal creep.*

 
www.witpress.com, ISSN 1743-355X (on-line)
doi:10.2495/BE100021

# 1 Introduction

Mixing of reactants in integrated micro fluidic systems is an important issue, due to the role played in chemical or biological analysis. A variety of micro mixers has been recently developed; Mansur et al. [1] and Hessel et al. [2] present a complete review of the use and development of active and passive micro mixers. The common conclusion of both reviews relies in the lower mixing rate present in micro mixers, because of the dominant viscous laminar flow conducing to slowly mixing rates. Different mechanisms and configurations have been implemented to achieve better mixing conditions, but the majority employ active mixers regardless of the simple fabrication technology and the easy implementation of passive mixers. Examples of active mixers include the swirl micro mixer [3], oscillating flat plate stirrer [4] and vortex pump mixer [5], while for passive mixers it is possible to find simpler versions, such as the Y and T-type flow [6] and more complex ones, such as chaotic micro mixers [7]. Active mixers could employ different stirrer shapes (concentric or eccentric symmetric and asymmetric rotors) to generate secondary flow, swirling flow and vortices, which enhance mixing performance [1].

When geometry devices are scaled down, the surface-to-volume ratio increases dramatically so that the surface related phenomena become increasingly dominant, e.g. micro heat exchangers and micro mixers present higher heat and mass transfer rates than macro systems of equal capacity [3]. Some new features emerge when mechanical structures are sufficiently small, and it becomes important to understand the various types of interactions that arise between the fluid flow constituents and the solid surfaces that contain them. For instance, a phenomenon known as the slip flow regime could emerges as the consequence of an insufficient number of molecules in the sampling region [8] or hydrophilic and hydrophobic recovering quality of surfaces in contact with fluids [9]. Neto et al. [10], in a review article, highlight the need of properly describing the flow near the fluid-solid interfaces, because of its relevance to a wide range of applications, from lubrication to micro fluidics.

In the case of micro fluid flow of rarefied gases, the thermal creep or transpiration also appears as a consequence of inequalities in temperatures, which forces the fluid to slide over a surface from colder to hotter regions [11]. Rarefied gas flows are generally associated with low-density conditions, such as high-altitude and vacuum. However, the small length scales commonly encountered in micro fluid flows imply that rarefaction effects will be important at much higher pressures, for more details see [12].

Taking into account the advantages of the BIMs for the numerical simulation of micro fluid flow under the slip condition, this work presents a numerical approach based on the use of the normal and tangential projections of the velocity integral representation formula for Stokes flows, resulting in weakly singular mixed system of integral equations of the first and second kinds for the normal and tangential components of the surface traction. The proposed approach is used to study flows in a Single rotor mixer (see Figure 1) under linear slip conditions with the thermal jump effect at the surface. The development of these formulations permits the inclusion of linear slip conditions into the boundary integral expressions directly, allowing the evaluation of micro flow in plane and curved geometries not subjected to symmetry conditions. The type of rotor mixer analyzed is of the single-shaft mixer type with an impeller rotating in close proximity to a stationary housing.

## 2 Governing equations

Fluid flow in micro scale devices usually happens at a very low Reynolds number due to their small characteristic length scale. In these cases, the flow of an incompressible Newtonian viscous fluid can be modelled by the Stokes system of equations, which states a balance between the pressure and the viscous-shear forces. In terms of a characteristic velocity $U$, a length scale $h$ and pressure $\mu\, U/h$, the Stokes system of equation can be written in dimensionless form as:

$$\frac{\partial u_i}{\partial x_i} = 0 \qquad \frac{\partial \sigma_{ij}}{\partial x_j} = 0 \tag{1}$$

where

$$\sigma_{ij} = -p\delta_{ij} + \left(\frac{\partial u_i}{\partial x_j} + \frac{\partial u_j}{\partial x_i}\right) \tag{2}$$

$\vec{u}$ being the velocity vector, $p$ the pressure and $\delta_{ij}$ the Kronecker delta.

Under these confined flow configurations, appropriate velocity slip conditions at the wall surfaces must be prescribed in order to employ continuum models for the description of the flow field. The Navier's slip boundary condition states that the relative tangential fluid velocity, $u_t^f$, with respect to the tangential wall velocity, $U_t^w$, is directly proportional to the tangential projection of the local shear rate, $\dot{\gamma}_t$, as given by equation (3). The proportional constant is called slip length, $L_s$ (dimensionless slip length in our case), and represents the hypothetical outward distance at the wall needed to satisfy the no-slip flow condition [12].

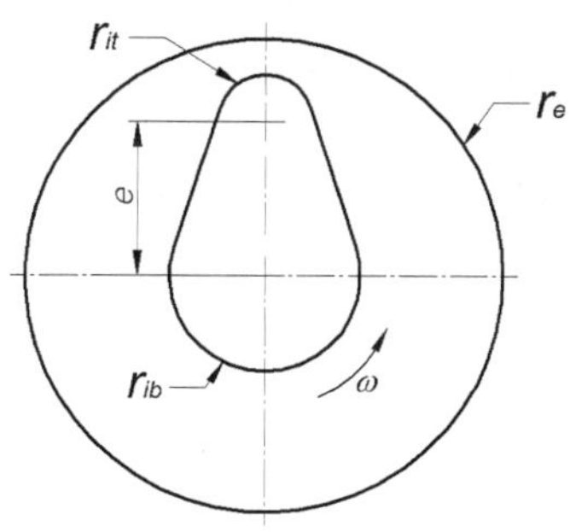


Figure 1: Single rotor mixer geometry.

Besides, when dealing with rarefied gases the effect of thermal creep or transpiration needs to be considered, as it relates the tangential fluid velocity with the tangential heat flux at the walls. In this way, the complete slip condition is described by:

$$u_t^f - U_t^w = L_s \dot{\gamma}_t + L_{sT}(-q_s) \tag{3}$$

with

$$\dot{\gamma}_t = \left(\frac{\partial u_i}{\partial x_j} + \frac{\partial u_j}{\partial x_i}\right) n_j s_i = \left(\frac{\partial u_i}{\partial n} s_i + \frac{\partial u_j}{\partial s} n_j\right) \tag{4}$$

where $n_i$ and $s_i$ are respectively the $i$ components of the normal and tangential vectors to a boundary surface and $q_s = -k\, \partial T/\partial s$ is the tangential heat flux at the boundary contours with $k$ as the thermal conductivity. In the above expression, $L_{sT}$ is a coefficient constant proportional to the product of the Reynolds number times the square of the Knudsen number and inversely proportional to the Eckert number.

For the implementation of this type of slip condition in the velocity integral representational formula for Stokes flows, the tangential shear rate at solid-fluid interfaces can be evaluated in terms of the surface traction force $\bar{f}$, which tangential projection can be expressed as:

$$\begin{aligned} f_t = f_i s_i = \sigma_{ij} n_j s_i &= \left[-p\delta_{ij} n_i s_i + \left(\frac{\partial u_i}{\partial x_j} + \frac{\partial u_j}{\partial x_i}\right) n_j s_i\right] \\ &= \left(\frac{\partial u_i}{\partial x_j} + \frac{\partial u_j}{\partial x_i}\right) n_j s_i = \dot{\gamma}_t \end{aligned} \tag{5}$$

where for consistence with our formulation, the above expression has been written in dimensionless form, with a characteristic traction force $\mu\, U/h$.

To complete the boundary conditions at the solid surfaces, the following non-flux condition across the boundaries needs to be considered:

$$u_i(x) n_i(x) = U_n^w(x), \ \text{at any points x on the boundary surfaces} \tag{6}$$

where the following integral relation needs to be satisfied, according to the conservation of mass for an incompressible fluid:

$$\int_{S = S_{int} \cup S_{ext}} U_n^w(x)\, dS_x = 0 \tag{7}$$

with $S$ as the union of the external and internal surfaces, $S_{ext}$ and $S_{int}$, respectively. In our case for micro mixers with external stationary housing, $U_n^w(x) = U_t^w(x) = 0$ at $x \in S_{ext}$.

## 3 Boundary integral formulation for slip Stokes flows

The Stokes velocity field has the following direct integral representation formulae for an arbitrary point $x$ in a closed domain $\Omega$ filled with a Newtonian fluid [13]:

$$c(x)u_i(x) - \int_S K_{ij}(x,y)u_j(y)dS_y + \int_S u_i^j(x,y)f_j(y)dS_y = 0 \tag{8}$$

where $c$ is a constant dependent on the position of the source point. For internal points $c = 1$, for point at a smooth boundaries $c = 1/2$ and for external points $c = 0$.

The kernels of the above integral representational formula are the Stokeslet and its corresponding surface tractions or Stresslet, which two dimension expressions are given by:

$$u_i^j(x,y) = -\frac{1}{4\pi}\left[ln\left(\frac{1}{r}\right)\delta_{ij} + \frac{(x_i - y_i)(x_j - y_j)}{r^2}\right] \tag{9a}$$

$$K_{ij}(x,y) = -\frac{1}{\pi}\frac{(x_i - y_i)(x_j - y_j)(x_k - y_k)n_k(y)}{r^4} \tag{9b}$$

$r$ being the Euclidean distance between point $x$ and $y$, i.e. $r = |x - y|$.

Substituting the above slip condition with consideration of thermal jump, equation (3), into the normal and tangential projections of the Stokes velocity field, the following system of integral equations for the normal and tangential components of the unknown surface traction is found:

$$\begin{aligned}
&\int_S u_i^j(x,y)\big(f_n(y)n_j(y) + f_t(y)s_j(y)\big)n_i(x)dS_y \\
&\qquad - \int_S K_{ij}(x,y)L_S\, f_t(y)s_j(y)n_i(x)dS_y = -c(x)U_n^w \\
&\qquad + \int_S K_{ij}(x,y)\big(U_j^w(y)\big)n_i(x)dS_y \\
&\qquad + \int_S K_{ij}(x,y)\left(L_{ST}\frac{\partial T(y)}{\partial s}s_j(y)\right)n_i(x)dS_y
\end{aligned} \tag{10}$$

and

$$
\begin{aligned}
c(x)L_S\, f_t(x) + \int_S u_i^j(x,y)\left(f_n(y)n_j(y) + f_t(y)s_j(y)\right)s_i(x)dS_y \\
- \int_S K_{ij}(x,y)L_S\ f_t(y)\ s_j(y)\ s_i(x)dS_y \\
= -c(x)U_t^w \\
+ \int_S K_{ij}(x,y)\left(U_j^w(y)\right)s_i(x)dS_y \\
+ \int_S K_{ij}(x,y)\left(L_{sT}\frac{\partial T(y)}{\partial s}s_j(y)\right)s_i(x)dS_y
\end{aligned}
\tag{11}
$$

The limiting value of the integral kernels in (10) and (11) as the radius $r$ tends to zero is given by:

$$
\begin{aligned}
K_{ij}(x,y)n_j(y)n_i(x) &= -\frac{r^2}{\pi}\left(\frac{1}{r}\frac{\partial r}{\partial n_y}\right)^2\left(\frac{1}{r}\frac{\partial r}{\partial n_x}\right) = -\frac{r^2}{\pi}\left(\frac{k(x)}{2}\right)^3 \to 0 \\
K_{ij}(x,y)s_j(y)n_i(x) = K_{ij}(x,y)n_j(y)s_i(x) &= -\frac{r}{\pi}\frac{\partial r}{\partial s_x}\left(\frac{1}{r}\frac{\partial r}{\partial n_y}\right)^2 \\
&= -\frac{r}{\pi}(\sin(\vec{n},\vec{r}))\left(\frac{k(x)}{2}\right)^2 \to 0 \\
u_i^j(x,y)n_j(y)s_i(x) = u_i^j(x,y)s_j(y)n_i(x) &= -\frac{r}{4\pi}\frac{\partial r}{\partial s_y}\left(\frac{1}{r}\frac{\partial r}{\partial n_x}\right) \\
&= -\frac{r}{4\pi}(\sin(\vec{n},\vec{r}))\left(\frac{k(x)}{2}\right) \to 0 \\
u_i^j(x,y)n_j(y)n_i(x) &= -\frac{1}{4\pi}\left[ln\left(\frac{1}{r}\right) + r^2\left(\frac{1}{r}\frac{\partial r}{\partial n_y}\right)\left(\frac{1}{r}\frac{\partial r}{\partial n_x}\right)\right] \\
&= -\frac{1}{4\pi}\left[ln\left(\frac{1}{r}\right) - r^2\left(\frac{k(x)}{2}\right)^2\right] \to -\frac{1}{4\pi}ln\left(\frac{1}{r}\right) \\
u_i^j(x,y)s_j(y)s_i(x) &= -\frac{1}{4\pi}\left[ln\left(\frac{1}{r}\right) + \frac{\partial r}{\partial s_y}\frac{\partial r}{\partial s_x}\right] = -\frac{1}{4\pi}\left[ln\left(\frac{1}{r}\right) - \cos^2(\vec{s},\vec{r})\right] \\
&= -\frac{1}{4\pi}\left[ln\left(\frac{1}{r}\right) - (1-\cos^2(\vec{n},\vec{r}))\right] \\
&= -\frac{1}{4\pi}\left[ln\left(\frac{1}{r}\right) - 1 + r^2\left(\frac{k(x)}{2}\right)^2\right] \to -\frac{1}{4\pi}\left(ln\left(\frac{1}{r}\right) - 1\right)
\end{aligned}
$$

with only logarithmic singularities on $u_i^j(x,y)n_j(y)n_i(x)$ and $u_i^j(x,y)s_j(y)s_i(x)$.

In the above limiting values, the term $\partial r/r\partial n$ tends to $k(x)/2$ as the radius $r$ tends to zero, where $k(x)$ is the curvature at a point $x$ on the boundaries (see Courant and Hilbert [14, Vol. 2, page 299]).

To solve equations (10) and (11), it is necessary to know the temperature field with the aim of obtaining the temperature gradient in the tangential direction at boundary contours. The temperature field is found by solving the energy equation, written here in dimensionless form as:

$$Pe\left(\frac{\partial T}{\partial t}+u_j\frac{\partial T}{\partial x_j}\right)=\frac{\partial^2 T}{\partial x_j^2} \tag{12}$$

where $Pe=(U^3\mu/\rho gk)^{1/2}$ is the Peclet number, which can be expressed in terms of the Reynolds number through the Prandtl number as $Pr=Pe/Re=\mu/\rho k$. For cases of $Pr=O(1)$, our previous assumption of small Reynolds number implies that $Pe\ll 1$, reducing the above energy equation (at the first order of approximation in $Pe$) to the Laplace equation, i.e. quasi-static approximation, the integral representation formula of which is:

$$c(x)T(x)=\int_S\frac{\partial\phi(x,y)}{\partial n}T(y)\,dS_y-\int_S\phi(x,y)\frac{\partial T(y)}{\partial n}dS_y \tag{13}$$

The kernels of the above integral representational formula are $\phi(r)=-\ln(r)/2\pi$ for the surface single layer potential and $\partial\phi(x,y)/\partial n=(x_i-y_i)n_i(y)/2\pi r^2=\cos(\vec{n},\vec{r})/2\pi r$ for the surface double layer potential, which are weak singular and regular, respectively (see our previous limiting values as $r$ tends to zero).

In the slip flow region, the temperature field is also subject to a jump condition at the boundary contours, [14], given by the radiation condition

$$T^f-T^w=L_T(-q_n) \tag{14}$$

with $q_n=-k\,\partial T/\partial n$ as the normal heat flux to the surface and $L_T$ as a coefficient constant function of the Knudsen and Prandtl numbers and the energy accommodation coefficient. The above jump condition represents a Robin type boundary condition for the temperature field. Substituting equation (14) into the integral representational formula for the temperature, equation (13), leads to the following boundary integral equation for the unknown normal flux:

$$\begin{aligned}c(x)L_T\frac{\partial T(x)}{\partial n}-\int_S\frac{\partial\phi(x,y)}{\partial n}L_T\frac{\partial T(y)}{\partial n}dS_y+\int_S\phi(x,y)\frac{\partial T(y)}{\partial n}dS_y\\ =-c(x)T^w(x)+\int_S\frac{\partial\phi(x,y)}{\partial n}T^w(y)\,dS_y\end{aligned} \tag{15}$$

The above integral equation can be solved by a classical BEM procedure; the quadratic scheme is the present work. The obtained surface heat flux can be substituted into the temperature jump condition (14) to find the corresponding fluid temperature at contact with the solid surfaces.

The tangential heat flux at the boundary contours is obtained by taking the limiting value of the gradient of the temperature field, at an internal point, approaching a boundary point, and then multiplying the resulting expression by the surface tangential vector. In this way, the surface gradient of the temperature field is given by:

$$c(x)\frac{\partial T(x)}{\partial x_j} = \int_S \frac{\partial \phi(x,y)}{\partial x_j}\frac{\partial T(y)}{\partial n} dS_y - \int_S \frac{\partial^2 \phi(x,y)}{\partial x_j \partial n} T(y)\, dS_y \tag{16}$$

where the first integral is a singular integral of the Cauchy type and the second a hyper singular integral. However, by multiplying the above relation by the surface tangential vector, the resulting integral relation for the tangential derivative only possesses singularities of the Cauchy type:

$$\begin{aligned} c(x)\frac{\partial T(x)}{\partial s} &= \int_S \frac{\partial \phi(x,y)}{\partial s}\frac{\partial T(y)}{\partial n} dS_y \\ &- \int_S \frac{\partial^2 \phi(x,y)}{\partial s \partial n} T(y)\, dS_y \end{aligned} \tag{17}$$

where the limiting value of the above kernels as the distance $r$ tends to zero is given by:

$$\frac{\partial \phi(x,y)}{\partial s} = -\frac{1}{2\pi}\frac{\partial ln(r)}{\partial x_j} s_j = -\frac{(x_j - y_j)}{2\pi\, r^2} s_j = \frac{1}{2\pi}\frac{sin(\vec{n},\vec{r})}{r} \rightarrow \frac{1}{r}$$

$$\begin{aligned} \frac{\partial^2 \phi(x,y)}{\partial s \partial n} &= \frac{\delta_{ij}}{2\pi r^2} n_j s_i - \frac{(x_j - y_j)(x_i - y_i)}{\pi\, r^4} n_j s_i \\ &= \frac{n_j(y) s_i(x)}{2\pi r^2} - \frac{1}{\pi}\frac{cos(\vec{n},\vec{r})}{r}\frac{sin(\vec{n},\vec{r})}{r} \rightarrow -\frac{1}{\pi}\frac{k(x)}{2}\frac{sin(\vec{n},\vec{r})}{r} \end{aligned}$$

The evaluation of the above integrals needs to be considered in the sense of Cauchy principal value. Various regularization methods to reduce the order of singularity of this type of integrals are available in the literature; most of them are based on expanding the kernel around the singular point and subtraction of the most singular part as originally suggested by Mikhlin [16]. A simple alternative is to find the integral representational formulae of a known potential field, with the same singularity as the field in consideration. Subtraction of both integral equations removes the most singular part of the integral operators (for more details see [16]). In general, this is not a simple task, but in the present case it is possible to define the potential

$$\tilde{T}(x) = \frac{\partial T(p)}{\partial x_j}\left(x_j - x(p)_j\right) + T(p), \tag{18}$$

solution of the Laplace equation $\nabla^2\tilde{T} = 0$, with values of the constants $T(p)$ and $\partial T(p)/\partial x_j$ equal to the values of the temperature field (13) and its gradient at a given evaluation point $\vec{x}(p)$. Therefore, at the evaluation point $\vec{x}(p)$, $\tilde{T}(p) = T(p)$ and $\frac{\partial \tilde{T}(p)}{\partial x_j} = \frac{\partial T(p)}{\partial x_j}$. By subtracting the integral representation formula of the tangential derivative of both fields, we obtain

$$c(x)\left(\frac{\partial T(x)}{\partial s} - \frac{\partial \tilde{T}(x)}{\partial s}\right) = \int_S \frac{\partial \phi(x,y)}{\partial s}\left(\frac{\partial T(y)}{\partial n} - \frac{\partial \tilde{T}(y)}{\partial n}\right) dS_y - \int_S \frac{\partial^2 \phi(x,y)}{\partial s \partial n}\left(T(y) - \tilde{T}(y)\right) dS_y$$

At an evaluation point $\vec{x}(p)$, the above integral relation reduces to the following regular integral equation:

$$\int_S \frac{\partial^2 \phi(p,y)}{\partial s \partial n}\left(T(y) - T(p) - \frac{\partial T(p)}{\partial x_j}(y_j - p_j)\right) dS_y = \int_S \frac{\partial \phi(p,y)}{\partial s}\left(\frac{\partial T(y)}{\partial n} - \frac{\partial T(p)}{\partial n}\right) dS_y \tag{19}$$

Given the values of the temperature $T$ and its normal derivative $\frac{\partial T}{\partial n}$, obtained from (14) and (15), equation (19) provides a linear relationship between the directional derivatives of the temperature at the evaluation point $\vec{x}(p)$, i.e. between the values of $\frac{\partial T(p)}{\partial x_1}$ and $\frac{\partial T(p)}{\partial x_2}$. A second relation between these two values is given by the known normal derivative at $\vec{x}(p)$, i.e. $\frac{\partial T(p)}{\partial n} = \frac{\partial T(p)}{\partial x_1} n_1 + \frac{\partial T(p)}{\partial x_2} n_2$. From these two expressions, it is possible to obtain the directional derivatives of the temperature field at each boundary point and find the corresponding tangential derivative $\frac{\partial T(p)}{\partial s} = \frac{\partial T(p)}{\partial x_1} s_1 + \frac{\partial T(p)}{\partial x_2} s_2$ to be used in the integral equations (10) and (11) for the normal and tangential components of the surface velocity in order to determine the corresponding effect of the thermal creep on the flow field.

Figure 2 shows the obtained fluid temperature distribution at the rotor and housing boundaries of the single rotor mixer given in Figure 1, for different constant temperature at the solid walls. Two different cases are reported in the figure, corresponding to a hotter and colder rotor, i.e. $T_{ri} > T_{re}$ and $T_{re} > T_{ri}$

respectively. In Figure 3, we report the corresponding velocity fields inside the mixer in the case of a stationary rotor, corresponding to only a thermal creep fluid motion. Two symmetric recirculation regions are formed around the rotor head, where the direction of the recirculation motion changes according to the direction of the heat flux, i.e. $T_{ri} > T_{re}$ or $T_{re} > T_{ri}$. Similar recirculation patterns are reported by Lockerby et al. [12] in the case of eccentric circular cylinders, however in their analysis they do not use a jump temperature condition, as in (13), but instead they use a high order thermal stress slip condition, where the tangential velocity is proportional to the tangential derivative of the normal heat flux.

In order to see the thermal creep effect on the recirculation pattern at the rotor head of a rotating mixer, in Figure 4 is reported the obtained flow fields corresponding to $L_{ST} = 0.1, 1$ and $10$. As can be observed from the figure, as the value of $L_{ST}$ increases the recirculation and the slip velocity at the rotor head are strongly affected by the thermal creep until the value of $L_{ST}$ becomes too large so that the thermal creep effect dominates the flow behaviour and a recirculation zone appears at the sides of the rotor head, as those observed in Figure 3 for the case of only thermal creep, but in the present case they are not located symmetrically. In addition, in the condition of very large value $L_{ST}$, the position of the recirculation zones changes with the direction of the heat flux.

## 4 Conclusions

In this work, the thermal creep effect at micro scale flows has been evaluated. This phenomenon usually takes place due to micro fluid flow of rarefied gases and could affect the velocity field due to a heat flux and a temperature difference in the domain. An integral equation approach based on the normal and tangential projections of the direct boundary integral representational formula for the Stokes velocity field is developed for the numerical simulation of creeping flow

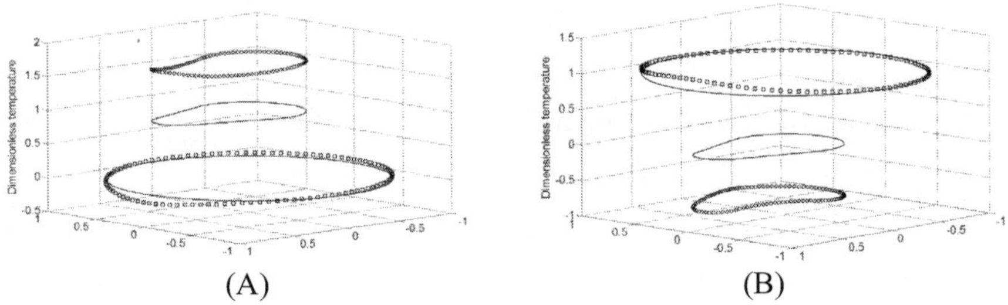


(A) (B)

Figure 2: Temperature profiles at single rotor boundaries for $L_T = 1$: $T_{ri} > T_{re}$ (A) and $T_{re} > T_{ri}$, (B). Solid lines represent the wall temperatures.

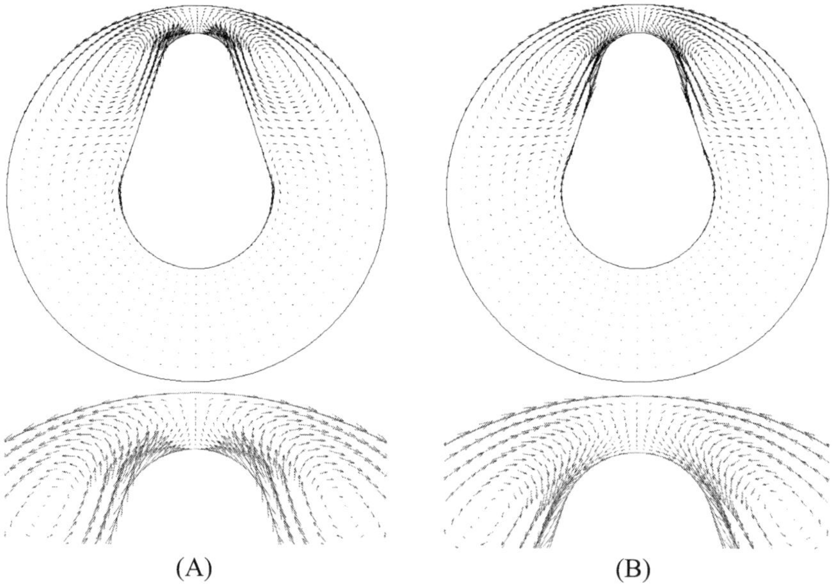

Figure 3: Vector velocities for $L_s$= 0.5, $L_T$ = 1 and $L_{sT}$ = 10 and $T_{ri} > T_{re}$ (A) and $T_{re} > T_{ri}$, (B), without considering rotational velocity at the inlet surface.

under linear slip boundary conditions and applied to analyze the performance of a single rotor mixer subjected to thermal creep. The slip condition was included in the boundary integral formulation by expressing the tangential shear rate in terms of the tangential component of the surface traction vector. The projection of the velocity integral representational formula on the normal and tangential directions smoothes the singularity of the integral kernels, resulting only in a weak singular kernel of the logarithm type, which can be numerically integrated by using Telles' transformation and standard Gaussian Quadrature formulae.

The integral formulation has been modified to account for the thermal effect, which is a function of the tangential heat flux. For the latter, an integral representational formulae was obtained by taking the limiting value of the gradient of the temperature field and multiplying the resulting relation by the surface tangential vector, presenting only singularities of the Cauchy type. The order of singularity of this type of integral was reduced by finding the integral representational formulae of a known potential field, with the same singularity as the field in consideration, reducing the expression to a regular integral equation at the evaluation point. The implementation of this formulation allowed us to observe that the thermal creep effect can modify the recirculation pattern at the rotor head of a rotating mixer dominating the flow behaviour and leading to recirculation zones that appear at different sides of the rotor head depending on the direction of the heat flux (see Figure 4).

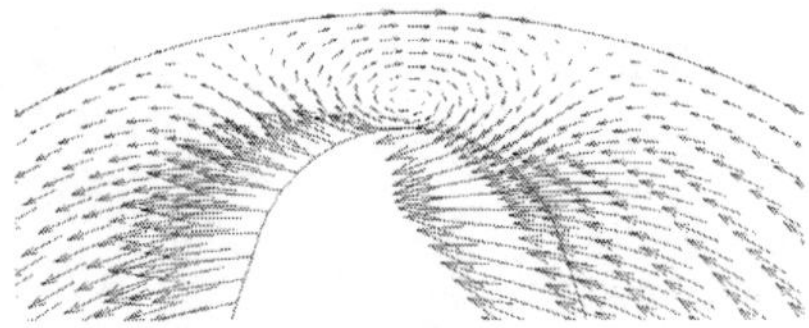

(A)

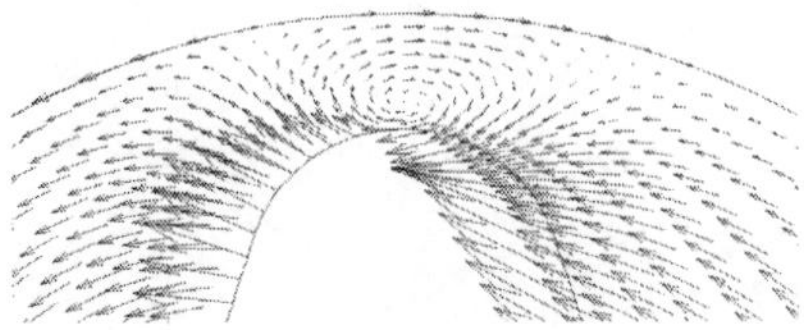

(B)

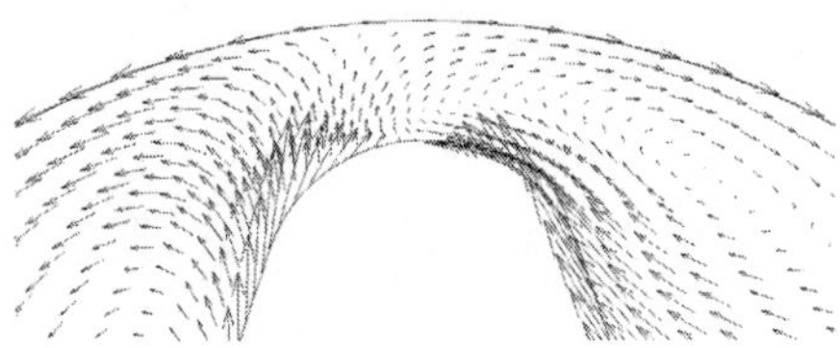

(C)

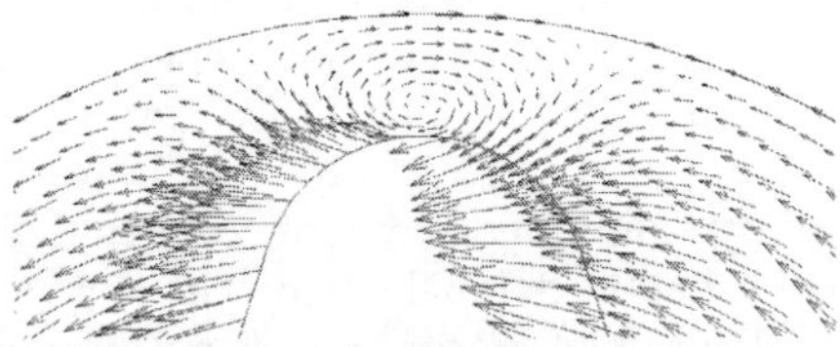

(D)

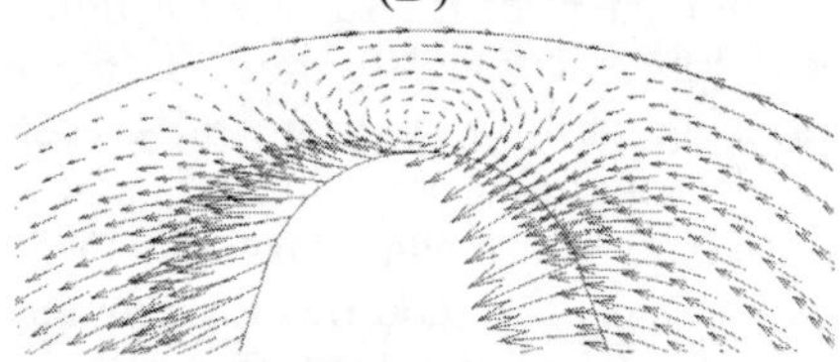

(E)

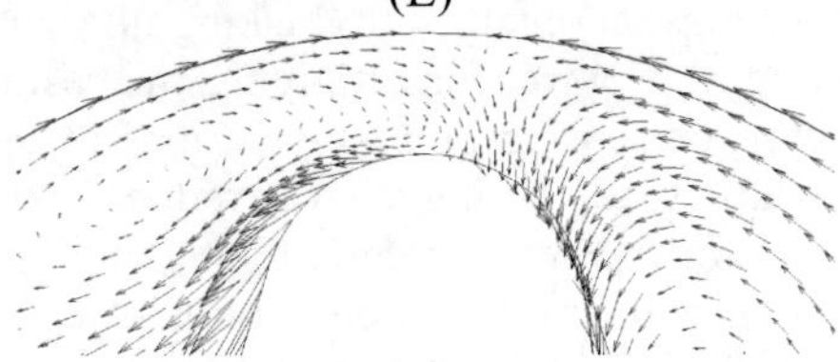

(F)

Figure 4: Vector velocities with $L_S$= 0.1 and $L_T$ = 1, $L_{ST}$ = 0.1 (A), $L_{ST}$ = 1 (B), $L_{ST}$ = 10 (C), $L_{ST}$ = 0.1 (D), $L_{ST}$ = 1 (E), $L_{ST}$ = 10 (F). Results in (A)-(C) are for $T_{ri} > T_{re}$, while in (D)-(F) are for $T_{re} > T_{ri}$.

## Acknowledgements

The first author acknowledges the support of the PhD program at the Universidad Pontificia Bolivariana and COLCIENCIAS, Colombia, under the support provided with the project "Desarrollo de geometrías para aplicación industrial en microintercambiadores de calor" code 1210-479-21999, contract 436-2008.

## References

[1] E.A. Mansur, Y. Mingxing, W. Yundong and D. Youyuan, *A state-of-the-art review of mixing in microfluidic mixers*, Chin. J. Chem. Eng., 16 (2008), 503–516.
[2] V. Hessel, H. Lwe, F. Schnfeld, *Micromixers: a review on passive and active mixing principles*, Chemical Engineering Science, 60 (2005), 2479–2501.
[3] G. Hu and D. Li, *Multiscale phenomena in microfluidics and nanofluidics*, Chemical Engineering Science, 62 (2007), 3443–3454.
[4] Jae-Yong Park, Yong-Dae Kim, Sang-Rak Kim, Seog-Young Han and Joo-SungMaeng, *Robust design of an active micro-mixer based on the Taguchi method*, Sensors and Actuators B, 129 (2008), 790–798.
[5] K.F. Lei and W.J. Li, *A novel in-plane microfluidic mixer using vortex pumps for fluidic discretization*, JALA, 13 (2008), 227–236.
[6] A. Goullet, I. Glasgow, and N. Aubry, *Effects of microchannel geometry on pulsed flow mixing*, Mech. Res. Commun., 33 (2006), 739–746.
[7] A.A. Mouza, C.-M. Patsa, F. Schnfeld, *Mixing performance of a chaotic micro-mixer*, Chemical Engineering Research and Design, 86 (2008), 1128–1134.
[8] P. A. Thompson and S. M. Troian, *A general boundary condition for liquid flow at solid surfaces*, Nature, 389 (1997), 360–362.
[9] O.I. Vinogradova, *Slippage of water over hydrophobic surfaces*, Int. Journal of Miner. Process., 56 (1999), 31–60.
[10] C. Neto, D.R. Evans, E. Bonaccurso, H. Butt and V. Craig, *Boundary slip in Newtonian liquids: a review of experimental studies*, Rep. Progr. Phys., 68 (2005), 2859–2897.
[11] G. Karniadakis, A. Beskok and N. Aluru, "Microflows and Nanoflows: Fundamentals and Simulation," 1st edition, Springer, New York, 2005.
[12] D. Lockerby, J. M. Reese, D. R. Emerson and R. W. Barber. *Velocity boundary condition at solid walls in rarefied gas calculations*, Physical Review E 70, (2004), 017303-1 - 017303-9.
[13] H. Power and L.C. Wrobel, "Boundary Integral Methods in Fluid Mechanics," 1st edition, Computational Mechanics Publications, Southampton, 1995.
[14] R. Courant and D. Hilbert, Methods of mathematical physics, 3rd edition, Cambridge University Press, Cambridge, 1962.

[15] O. Aydın, M, Avcı. *Heat and fluid flow characteristics of gases in micropipes.* International Journal of Heat and Mass Transfer, 49, (2006), 1723–1730.
[16] S.G. Mikhlin, Multidimensional Singular Integrals and Integral Equations. Pergamon Press, New York, 1957.
[17] H. Chen, J. Jin, P. Zhang and P. Lu, *Multi-variable non-singular BEM for 2-D potential problems*, Tsinghua Science and Technology, Vol 10, 1, 43-50.

# Solution of velocity-vorticity URANS by BEM

L. Škerget & J. Ravnik
*Faculty of Mechanical Engineering, University of Maribor, Slovenia*

## Abstract

This paper describes the fundamental aspects of turbulence modelling for incompressible fluid flow and corresponding numerical models based on the boundary-domain integral equations. The velocity-vorticity formulation of the Reynolds averaged Navier-Stokes (RANS) equations is used, while the averaged pressure field is determined by solving the Poisson velocity equation. Several low-Reynolds-number turbulent models, e.g. Launder-Sharma, are applied.

## 1 Introduction

The set of partial differential equations governing the motion of viscous fluid is known as nonlinear Navier-Stokes equations. This equation system is generally considered to be the fundamental description for all laminar as well as turbulent flows, although some statistical averaging procedure is required in practice to predict the turbulence and simulate numerically the flow at higher Reynolds number values due to the enormous computational effort needed. Turbulence is a highly complex transport phenomenon. A number of well established characteristics of a turbulent motion can be summarized as: highly nonlinear transport process, highly diffusive flow, three-dimensional flow, flow with multiple length and time scales, and time dependent (stochastic) transport phenomenon with identifiable statistical flow properties. In most cases of modelling and simulation of real engineering turbulent flows we are not interested in detailed resolution of all turbulent flow scales in favour of some type of averaged flow description. Therefore, turbulence effects enter the flow description via a model that is typically based on a combination of theory and experiment.

Various approaches exist for the turbulent flow prediction, i.e., full turbulence simulation, large eddy simulation, Reynolds averaged models, etc [1]. However, the extremely small time and length scales associated with turbulent fluid motion

WIT Transactions on Modelling and Simulation, Vol 50, © 2010 WIT Press
www.witpress.com, ISSN 1743-355X (on-line)
doi:10.2495/BE100031

require impractically dense computational meshes and time steps when attempting to solve the unaveraged unsteady governing equations for practical engineering problems at realistic Reynolds number values.

The averaged form of the Navier-Stokes equations through the Reynolds decomposition of instantaneous value of each flow variable into a time averaged mean value, and an instantaneous deviation or fluctuation from the mean value is still the most commonly used approach in the numerical simulation of the engineering fluid flow problems with turbulence. The Reynolds mean flow equations involve more unknowns than equations. The term turbulence model implies assumptions, closing the mathematical description of the problem, i.e. makes the number of unknowns equal to the number of equations and enables the simulation of the flow in its important aspects. However, the computation of the turbulent flow based on any given turbulent model generally requires a knowledge of empirical constants.

## 2 Governing equations

### 2.1 Primitive variables formulation

When considering the turbulent flow a time averaged form of the Navier-Stokes equations is usually employed through the Reynolds decomposition of instantaneous value of each flow variable, e.g. velocity vector $v_i(r_j,t)$ into a time-averaged mean value $\overline{v_i}$ and an instantaneous deviation or fluctuation $v_i'$ from the mean value, such that $v_i = \overline{v_i} + v_i'$, and similarly pressure $p(r_j,t)$ into a time-averaged mean value $\overline{p}$ and an instantaneous deviation or fluctuation $p'$ from the mean value, such that $p = \overline{p} + p'$.

With the assumption of incompressibility the conservation equations set for the time-averaged mean field functions in a turbulent incompressible fluid motion, can be written as

$$\frac{\partial \overline{v_j}}{\partial x_j} = 0, \tag{1}$$

$$\rho_o \frac{D\overline{v_i}}{Dt} = -\frac{\partial \overline{p}}{\partial x_i} - \frac{\partial \overline{\tau_{ij}}}{\partial x_j} + \rho_o g_i, \tag{2}$$

for $i,j = 1,2,3$.

The time-averaged mean momentum flux tensor $\overline{\tau_{ij}}$ can be written as a sum of molecular diffusion flux for the time mean values and the turbulent flux, e.g. in the form

$$\overline{\tau_{ij}} = \overline{\tau_{ij}}^m + \overline{\tau_{ij}}^t = -2\eta_o \overline{\dot{\varepsilon}_{ij}} + \rho_o \overline{v_i' v_j'}, \tag{3}$$

in which the tensor quantity $\overline{\dot{\varepsilon}_{ij}}$ is the symmetrical part of the time-averaged mean velocity gradient

$$\overline{\dot{\varepsilon}_{ij}} = \frac{1}{2}\left(\frac{\partial \overline{v_i}}{\partial x_j} + \frac{\partial \overline{v_j}}{\partial x_i}\right). \tag{4}$$

The Reynolds averaged Navier-Stokes (RANS) equations set, governing the transport of time mean flow quantities in a turbulent incompressible fluid flow, can be obtained substituting eq.(3) in eq.(2), as follows

$$\frac{\partial \overline{v_j}}{\partial x_j} = 0, \tag{5}$$

$$\rho_o \frac{D\overline{v_i}}{Dt} = -\frac{\partial \overline{p}}{\partial x_i} + \frac{\partial}{\partial x_j}\left(2\eta_o \overline{\dot{\varepsilon}_{ij}}\right) + \rho_o g_i - \frac{\partial \rho_o \overline{v_i' v_j'}}{\partial x_j}, \tag{6}$$

With the exception of the additional Reynolds stress tensor or turbulent momentum flux tensor $\tau_{ij}^t$, the turbulent heat flux vector $q_i^t$, and the turbulent mass flux vector $j_i^t$

$$\tau_{ij}^t = \rho_o \overline{v_i' v_j'} = \rho_o \overline{\vec{v}' \otimes \vec{v}'}, \tag{7}$$

the mean flow quantities of a turbulent flow and the instantaneous flow quantities of a laminar flow satisfy the same set of differential equations; the no-slip condition on the solid surfaces and the free-stream condition for external flows, satisfied by the instantaneous velocity of the laminar flow, are also satisfied by the mean velocity of a turbulent flow.

## 2.2 Turbulence models

The completion of the mean flow governing equations requires that additional constitutive relations for the Reynolds fluxes be provided. This specification constitutes a turbulence model. An enormous variety of turbulence models exist, ranging in complexity from simple algebraic relations to descriptions involving multiple, nonlinear governing partial differential transport equations for specific turbulent quantities [2–4].

As noted previously, the term $\rho_o \overline{v_i' v_j'}$ appearing on the right hand side of eq.(6) is due to turbulent motion and is called the Reynolds stress. It acts to promote the diffusion of momentum and cannot be obtained by further time-averaging alone and has to be modelled. The main aim of all turbulence models is to find a rational closure of equations; any turbulent model requires a knowledge of empirical constants.

By analogy with the molecular momentum diffusion, the Boussinesq approximation expressed the turbulent momentum transport in terms of mean velocity gradients

$$\tau_{ij}^t = \rho_o \overline{v_i' v_j'} = -2\eta_t \overline{\dot{\varepsilon}_{ij}} + \frac{2}{3}\rho_o k \delta_{ij} = -\rho_o \nu_t \left(\frac{\partial \overline{v}_i}{\partial x_j} + \frac{\partial \overline{v}_j}{\partial x_i}\right) + \frac{2}{3}\rho_o k \delta_{ij}, \tag{8}$$

where $\nu_t = \eta_t/\rho_o$ is the eddy-viscosity or turbulent viscosity computed by the chosen turbulence model, and $k$ mean turbulent kinetic energy. Unlike the molecular momentum diffusivity coefficient $\nu$ which is a fluid property, the turbulent (eddy) diffusivity coefficient $\nu_t$ is local property of turbulent flow. Models for the turbulent transport properties attempt to produce reasonable turbulent effects within the governing equations as functions of the mean flow quantities.

### 2.3 Reynolds averaged Navier-Stokes mean flow equations

When the definitions in eq.(8) are substituted into governing eq.(6) the equations for the mean flow can be written in terms of effective or total momentum diffusivity $\nu_{ef} = \nu + \nu_t$, as follows

$$\frac{\partial \overline{v_j}}{\partial x_j} = 0, \tag{9}$$

$$\rho_o \frac{D\overline{v_i}}{Dt} = -\frac{\partial \overline{p}^\star}{\partial x_i} + \frac{\partial}{\partial x_j}\left(\rho_o 2\nu_{ef}\overline{\dot{\varepsilon}_{ij}}\right) + \rho_o g_i, \tag{10}$$

where the volumetric part of Reynolds stress has been included in the pressure term such as

$$\overline{p}^\star = \overline{p} + \frac{2}{3}\rho_o k. \tag{11}$$

The momentum equation eq.(10) can be written in a extended form suitable for velocity-vorticity formulation, e.g. in Cartesian tensor notation formulation

$$\rho_o \frac{D\overline{v_i}}{Dt} = -e_{ijk}\frac{\partial \overline{\omega_k}}{\partial x_j}\eta_{ef} + e_{ijk}\frac{\partial \eta_{ef}}{\partial x_j}\overline{\omega_k} + 2\frac{\partial \eta_{ef}}{\partial x_j}\frac{\partial \overline{v_i}}{\partial x_j} - \frac{\partial \overline{p}^\star}{\partial x_i} + \rho_o g_i\,. \tag{12}$$

Representing the effective transport properties, e.g. the effective dynamic viscosity $\eta_{ef}$ as a sum of a constant and variable part

$$\eta_{ef} = \eta_{efo} + \widetilde{\eta_{ef}}\,, \tag{13}$$

then the momentum eq.(12) can be written in analogy to the basic conservation equations formulated for the constant material properties

$$\rho_o \frac{D\overline{\vec{v}}}{Dt} = -\eta_{efo}\mathrm{rot}\overline{\vec{\omega}} - \mathrm{grad}\overline{p}^\star + \rho_o\vec{g} + \overline{\vec{f}^m}\,, \tag{14}$$

where the pseudo body force term $\overline{\vec{f}^m}$ is introduced into the momentum eq.(14) capturing the variable transport property effects, and given by expression

$$\overline{f_i^m} = -e_{ijk}\frac{\partial \overline{\omega_k}}{\partial x_j}\widetilde{\eta_{ef}} + e_{ijk}\frac{\partial \eta_{ef}}{\partial x_j}\overline{\omega_k} + 2\frac{\partial \eta_{ef}}{\partial x_j}\frac{\partial \overline{v_i}}{\partial x_j}\,, \tag{15}$$

which for plane flow problems reduces to relation

$$\overline{f_i^m} = -e_{ij}\frac{\partial \overline{\omega}}{\partial x_j}\widetilde{\eta_{ef}} + e_{ij}\frac{\partial \eta_{ef}}{\partial x_j}\overline{\omega} + 2\frac{\partial \eta_{ef}}{\partial x_j}\frac{\partial \overline{v_i}}{\partial x_j}\,. \tag{16}$$

Once the form of the eddy diffusivity coefficients are specified then the mean transport equations can be solved in the same manner as a laminar flow since the equations are the same except for augmented diffusivity coefficients. Though the turbulent flow problem has been reduced to a familiar system of partial differential transport equations, there remains the nontrivial task of determining how the eddy diffusivity coefficients vary with the flow field.

### 2.4 Velocity-vorticity mean flow formulation

With the mean vorticity vector $\overline{\omega_i}$ representing the curl of the mean velocity field $\overline{v_i}$

$$\overline{\omega_i} = e_{ijk}\frac{\partial \overline{v_k}}{\partial x_j} \qquad \text{and} \qquad \frac{\partial \overline{\omega_j}}{\partial x_j} = 0\,, \tag{17}$$

the fluid motion computation scheme is partitioned into its kinematics, given by the elliptic mean velocity vector equation

$$\frac{\partial^2 \overline{v_i}}{\partial x_j \partial x_j} + e_{ijk}\frac{\partial \overline{\omega_k}}{\partial x_j} = 0\,, \tag{18}$$

and kinetics given by mean vorticity transport equation, obtained as a curl of the mean momentum eq.(16), e.g., written in Cartesian tensor notation formulation

$$\frac{\partial \overline{\omega_i}}{\partial t} + \frac{\partial \overline{v_j}\,\overline{\omega_i}}{\partial x_j} = \nu_{efo}\frac{\partial^2 \overline{\omega_i}}{\partial x_j \partial x_j} + \frac{\partial \overline{\omega_j}\,\overline{v_i}}{\partial x_j} + \frac{1}{\rho_o} e_{ijk}\frac{\partial \overline{f_k^m}}{\partial x_j}\,, \tag{19}$$

which reduces for two-dimensional plane flow case, to the following scalar mean vorticity statement

$$\frac{\partial \overline{\omega}}{\partial t} + \frac{\partial \overline{v_j}\,\overline{\omega}}{\partial x_j} = \nu_{efo}\frac{\partial^2 \overline{\omega}}{\partial x_j \partial x_j} - \frac{1}{\rho_o} e_{ij}\frac{\partial \overline{f_i^m}}{\partial x_j}\,. \tag{20}$$

The vorticity transport eq.(19) is highly nonlinear partial differential equation. Due to the buoyancy force and variable effective transport properties, acting as additional nonlinear vorticity source terms, the vorticity transport equation is coupled to the energy/mass and transport equations for the turbulence quantities, i.e., '$k$' and '$e$' transport equations, making the numerical computation procedure very challenging.

## 3 Eddy-viscosity turbulence models

### 3.1 Two-equation LRN $k - \epsilon$ turbulence models

Low-Reynolds-number form turbulence models [5–7] seem to be necessary for accurate prediction of unsteady turbulent wall boundary flows where the near-wall physics is of interest. In an unsteady turbulent wall bounded flow, the near-wall region is characterized by a rapid phase change in flow quantities. The phase shift in the wall shear stress is important in engineering applications. Since at high frequencies the thickness of the unsteady layer can be of the same order of the viscous sublayer, incorrect near-wall asymptotic behaviour of a turbulence model may result in inaccurate prediction of near-wall unsteady turbulent stresses. In the vicinity of the wall, the presence of the solid wall causes preferential damping of the normal component of turbulent velocity. At the same time, the turbulence motion is restricted and the Reynolds number becomes small.

In the $k - \epsilon$ turbulence models, the turbulent motion is characterised by two quantities, namely the turbulent kinetic energy $k$ and the turbulent energy dissipation rate $\epsilon$, which extracts energy from the turbulence and converts this energy to heat. These values are given by relations

$$k = \frac{1}{2}\overline{v_i' v_i'}, \tag{21}$$

$$\epsilon = \nu \overline{\frac{\partial v_i'}{\partial x_j} \frac{\partial v_i'}{\partial x_j}}, \tag{22}$$

while the turbulent viscosity is given by the Kolmogorov-Prandtl relation

$$\eta_t = C_\eta f_\eta \frac{k^2}{\epsilon}, \tag{23}$$

which relates the eddy viscosity directly to the turbulence variables, $k$ and $\epsilon$, and where $C_\eta = 0.09$ is an empirical constant. The function $f_\eta$ is needed to interface the near-wall low-Reynolds-number eddy viscosity on the high-Reynolds-number eddy viscosity.

Both turbulent quantities $k$ and $\epsilon$ are determined from individual transport equations. The equations have also to be modelled, which means that several assumptions have to be considered in order to close the mathematical description, since these equations involve moments of fluctuating velocity of order higher than two. Since the turbulent fluctuations are damped near a solid wall, there is a region close to the wall where viscous effects are important. In this region, the local turbulence Reynolds number values, defined by the expressions

$$Re_t = \frac{\rho k^2}{\eta \epsilon}, \qquad Re_k = \frac{\rho \sqrt{k}\, y}{\eta},$$

$$Re_\tau = \frac{\rho v_\tau y}{\eta} = y^+, \qquad Re_K = \frac{\rho v_K y}{\eta} = y^*, \tag{24}$$

where $\eta$ is the molecular dynamic viscosity of the fluid, $y^+$ and $y^*$ are the dimensionless normal distances from wall based on the wall friction velocity $v_\tau = \sqrt{\tau_w/\rho} = \sqrt{\nu|\omega|}$ and Kolmogorov velocity scale $v_K = (\nu\epsilon)^{1/4}$, respectively, are small and the high Reynolds number turbulence models are no longer applicable. For this reason, low-Reynolds-number (LRN) $k - \epsilon$ turbulence models can be employed allowing integration right down to the wall with appropriate boundary conditions. This requires more grid nodes in the near-wall region than the other two schemes described above, but does permit a continuous solution without the problem of specifying an artificial match node. The damping of turbulence near a solid wall due to molecular diffusion is simulated through some damping functions attached to various terms of transport equations for the specific turbulent quantities in the region which allow a smooth change of the flow variable from the small viscous sublayer values very near the wall to the fully turbulent value away from the wall.

In the two equation $k-\epsilon$ low-Reynolds-number models with linear eddy-viscosity models, the values of $k$ and $\epsilon$ come directly from the differential transport equations for the turbulent kinetic energy and the eddy dissipation rate

$$\rho_o \frac{Dk}{Dt} = \frac{\partial}{\partial x_j}\left[\left(\eta_o + \frac{\eta_t}{\sigma_k}\right)\frac{\partial k}{\partial x_j}\right] + P_k - \rho_o\widetilde{\epsilon} - \rho_o D, \tag{25}$$

$$\rho_o \frac{D\widetilde{\epsilon}}{Dt} = \frac{\partial}{\partial x_j}\left[\left(\eta_o + \frac{\eta_t}{\sigma_\epsilon}\right)\frac{\partial \widetilde{\epsilon}}{\partial x_j}\right] + C_{\epsilon 1} f_{\epsilon 1}\frac{\widetilde{\epsilon}}{k}P_k - C_{\epsilon 2} f_{\epsilon 2}\rho_o\frac{\widetilde{\epsilon}^2}{k} + \rho_o E, \tag{26}$$

where the new model variable $\widetilde{\epsilon}$ introduced by Launder and Sharma [8] is defined as

$$\epsilon \equiv \widetilde{\epsilon} + D \qquad \text{with} \qquad D = 2.0\nu_o\left(\frac{\partial k^{1/2}}{\partial x_j}\right)^2. \tag{27}$$

Since the dissipation rate at the wall is equal to

$$\epsilon|_{wall} \equiv D|_{wall} = 2.0\nu_o\left(\frac{\partial k^{1/2}}{\partial x_j}\right)^2\bigg|_{wall}, \tag{28}$$

the variable $\widetilde{\epsilon}$ is zero at the wall which simplifies the specification of wall boundary conditions, i.e.

$$k = \widetilde{\epsilon} = 0\,. \tag{29}$$

Further, such definition of new variable $\widetilde{\epsilon}$ also gives rise to an extra term $E$ in the $\widetilde{\epsilon}$ transport equation

$$E = 2.0\nu_o\nu_t\left(\frac{\partial^2\overline{v_i}}{\partial x_j\partial x_k}\right)^2. \tag{30}$$

The turbulent kinetic energy production term $P_k$ is due to viscous forces and is modeled, e.g. by the following relation

$$P_k = 2\eta_t\overline{\dot{\varepsilon}_{ij}}\frac{\partial\overline{v_i}}{\partial x_j} = \eta_t\left(\frac{\partial\overline{v_i}}{\partial x_j} + \frac{\partial\overline{v_j}}{\partial x_i}\right)\frac{\partial\overline{v_i}}{\partial x_j}. \tag{31}$$

One of the most popular and well tested LRN $k-\widetilde{\epsilon}$ is the Launder and Sharma model (LS), where the damping functions are expressed as functions of the local turbulence Reynolds number as follows:

$$f_\eta = \exp\left[-\frac{3.4}{(1+0.02Re_t)^2}\right], \tag{32}$$

$$f_{\epsilon 1} = 1 \qquad \text{and} \qquad f_{\epsilon 2} = 1.00 - 0.3\exp\left(-Re_t^2\right), \tag{33}$$

with $C_{\epsilon 1} = 1.44$, $C_{\epsilon 2} = 1.92$, $\sigma_k = 1.0$, and $\sigma_\epsilon = 1.3$, while $\sigma_\rho = 0.9$, respectively.

The two equation, $k-\epsilon$ turbulence model described by eqs.(25) and (26) can be used in conjunction with the mean flow equations and the definition of $\eta_t$ given by eq.(23), to arrive at a continuum description of turbulent motion. This system of equations is highly nonlinear, with a strong coupling among various transport equations. Although, the model is far from being universal and has a number of weaknesses, it remains one of the most frequently used for real turbulent flow computation. It is also evident, that the size of the computational problem and numerical difficulties for a two equation turbulent simulation increases substantially over its laminar counterpart.

## 4 Numerical aspects: iterative strategy

Once a turbulence model has been selected for a specific application, it is important to consider how a numerical model, e.g. boundary element method, would be applied to produce correct flow simulation. The boundary element implementation of the two equation $k-\epsilon$ low-Reynolds-number model is basically straightforward since the transport equations for the turbulent quantities are of the familiar diffusion-convection type. The boundary element solution of these types of equations and of the velocity-vorticity formulation of Navier-Stokes equations was presented in [9–12] .

The discretised forms of eqs.(25) and (26) are highly nonlinear and strongly coupled to the mean flow RANS equations. It is also evident that the computational effort for a two equation turbulent simulation has increased substantially over its laminar counterpart. At the same time, the numerical difficulties have also increased significantly mainly due to two reasons: the nonlinear behaviour of the transport equations and boundary conditions for the turbulent transport and RANS partial differential equations. In particular, the dissipation equation may cause instabilities that lead to poor or nonconvergence of the numerical solution procedure. Stability problems can result in the prediction of negative values for both turbulent quantities $k$ and $\epsilon$. This non-physical behaviour of turbulent field functions can be also caused by a inadequate modelling of the source terms for $k$ and $\epsilon$. In many cases, the problem can be overcome by the clipping procedure in which negative values are replaced by small positive values.

Further, due to very sensitive nature of the $k$ and $\epsilon$ transport equations, it is of main importance to establish stable solution procedure. Generally, Picard or simple under-relaxation iteration technique should be applied in order to solve these coupled equations. The convergence of the $k$ and $\epsilon$ field functions is in general very slow and may not be monotonic. Though the convergence rate may be improved by the linearization of the $k$ and $\epsilon$ transport equations.

The solution iterative strategy is to solve for velocity $\vec{v}$ and vorticity $\vec{\omega}$ field functions and then to solve for $k$ and $\epsilon$ until convergence, repeating the iterative process if needed. The solution scheme is as follows:

1. Solve the Reynolds averaged Navier-Stokes equations
2. Solve for $k$ and $\epsilon$
    - 2.1 Update production $P_k$ and $P_\epsilon$
    - 2.2 Solve for $k$ - (until convergence)

2.3 Solve for $\epsilon$ - (until convergence)
2.4 Check convergence in terms of $\nu_t$. If not, go to 2.2.
3. Check convergence for $\vec{\omega}$. If not, go to 1.

For the given velocity field $\vec{v}$ the $k$ and $\epsilon$ equations have to be solved. The first point of the iterative scheme is that the equations for $k$ and $\epsilon$ are coupled iteratively. Therefore, the nonlinear transport equation is solved for $k$ first assuming that $\epsilon$ is known and then the computed value of $k$ is used to solve the nonlinear transport equation for $\epsilon$. Since it is not desired to deal with problems with a negative production term, the term $P_k$ is kept constant in eq.(25) and also production term $P_\epsilon$, i.e.

$$P_\epsilon = C_{\epsilon 1} f_{\epsilon 1} \frac{\epsilon}{k} P_k = C_{\epsilon 1} f_{\epsilon 1} \frac{P_k}{\tau_t}, \tag{34}$$

is kept constant in eq.(26).

Let us first consider eq.(25). In order to avoid keeping $\epsilon$ constant, the dissipation term is written as

$$D_k = \epsilon = \frac{C_\eta f_\eta}{\nu_t} k^2 . \tag{35}$$

Therefore, in the innermost iterative loop the linearized $k$ equation is solved, e.g. for given $\nu_t$ and $P_k$ solve until converge

$$\rho_o \frac{Dk_i}{Dt} = \frac{\partial}{\partial x_j}\left[\left(\eta_o + \frac{\eta_t}{\sigma_k}\right)\frac{\partial k_i}{\partial x_j}\right] + P_k - \rho_o \frac{C_\eta f_\eta}{\nu_t}\left(2k_{i-1}k_i - k_{i-1}^2\right), \tag{36}$$

where the subscript stands for the iteration counter.

Similarly, the following linearized $\epsilon$ equation is solved, e.g. for given $\nu_t$ $k$, and $P_\epsilon$ solve until converge

$$\rho_o \frac{D\epsilon_i}{Dt} = \frac{\partial}{\partial x_j}\left[\left(\eta_o + \frac{\eta_t}{\sigma_\epsilon}\right)\frac{\partial \epsilon_i}{\partial x_j}\right] + P_\epsilon - \rho_o \frac{C_{\epsilon 2} f_{\epsilon 2}}{k}\left(2\epsilon_{i-1}\epsilon_i - \epsilon_{i-1}\epsilon_{i-1}\right). \tag{37}$$

It can be observed, that linearized $k$ and $\epsilon$ eqs.(36) and (37) have the same structure. Now, the iterative scheme can be written in detail, as follows:

1. Compute $P_k$ and $P_\epsilon$
2. Repeat until convergence for $\nu_t$
   2.1 Solve for $k$:
      2.1.1 Solve eq.(25)
      2.1.2 $k_i = ur \cdot k_i + (1 - ur) \cdot k_{i-1}$
   2.2 Check convergence for $k$. If not, go to 2.1.
   2.3 Update $\nu_t$ using eq.(23)
   2.4 Solve for $\epsilon$:
      2.4.1 Solve eq.(26)
      2.4.2 $\epsilon_i = ur \cdot \epsilon_i + (1 - ur) \cdot \epsilon_{i-1}$
   2.4 Check convergence for $\epsilon$. If not, go to 2.4.
   2.5 Update $\nu_t$ using eq.(23)
3. Check convergence for $\nu_t$. If not, go to 2.

In monitoring the convergence of a turbulent flow simulation it is noteworthy that the velocity and vorticity field functions converge first. The grading of the computational mesh may have also a big impact on the solution stability, e.g., abrupt changes in mesh density can lead to divergence of the $k$ and $\epsilon$ solution process.

The discretization in generally greatly affects the numerical results, and this is especially true for the low-Reynolds turbulent models. The near-wall region where large gradients of the field functions occur has to be adequately discretized; the first internal nodes have to be placed into the viscous sublayer, e.g., of the order $y^+ = 0.5$ from the solid wall.

## 5 Boundary conditions for turbulent flows

In order to solve the complete dynamical coupled nonlinear equations system, some physically justified boundary conditions must be specified. Especially troublesome in this regard are $k$ and $\epsilon$ values for inlet boundaries, since to a large extent these values will set the turbulence level for the entire flow field. In general, boundary conditions assigned to diffusion-convective transport equations are of the first, second, mixed and outflow-convective type, imposed on the part of the boundary.

The boundary conditions applied at the specific parts of the surface are as follows, e.g., specified Dirichlet boundary conditions for the variables $(\overline{v_i}, \overline{k}, \overline{\widetilde{\epsilon}})$ at the inflow boundaries

$$\overline{v_i} = \overline{\overline{v}}_i, \qquad k = \overline{k} \qquad \text{and} \qquad \widetilde{\epsilon} = \overline{\widetilde{\epsilon}}, \tag{38}$$

zero Dirichlet values for the functions $(\overline{v_i}, \overline{k}, \overline{\widetilde{\epsilon}})$ at the solid wall boundaries, no-slip conditions,

$$\overline{v_i} = \overline{k} = \overline{\widetilde{\epsilon}} = 0, \tag{39}$$

or zero Dirichlet values for the functions $(\overline{v_i}, \overline{k})$ and zero Neumann for $\overline{\epsilon}$ at the solid wall boundaries

$$\overline{v_i} = \overline{k} = 0 \qquad \text{and} \qquad \overline{\frac{\partial \epsilon}{\partial n}} = 0, \tag{40}$$

and the outflow boundary conditions for the field quantities $(k, \widetilde{\epsilon})$ as zero Neumann normal flux values

$$\frac{\partial k}{\partial n} = \frac{\partial \widetilde{\epsilon}}{\partial n} = 0, \tag{41}$$

or in the form of outflow-convective boundary conditions

$$\frac{\partial k}{\partial t} + \langle \overline{v} \rangle \frac{\partial k}{\partial n} = 0 \qquad \text{and} \qquad \frac{\partial \widetilde{\epsilon}}{\partial t} + \langle \overline{v} \rangle \frac{\partial \widetilde{\epsilon}}{\partial n} = 0, \tag{42}$$

where $\langle \overline{v} \rangle$ is the mean outflow velocity.

## 6 Conclusion

In this work a numerical procedure based on the boundary element method for the simulation of two-dimensional is presented. Low Reynolds number $k-\epsilon$ has been considered. Iterative strategy of highly nonlinear and coupled governing equations is discussed. Future work will include two standard examples, e.g. fully developed turbulent flow [13–15] in channel and turbulent backward facing step flow [16], to test the proposed numerical algorithm. The nonlinearity of low Reynolds models is very severe and special attention will have to be paid to obtain convergence numerical solution.

## References


[1] Biswas, G. & Eswaran, V., *Turbulent Flows: Fundamentals, Experiments and Modeling*. Alpha Science International Ltd., Pangbourne, United Kingdom, 2002.

[2] Nagano, Y. & Hishida, M., Improved Form of the $k-\epsilon$ Model for Wall Turbulent Shear Flows. *ASME Journal of Fluids Engineering*, **109**, pp. 156–160, 1987.

[3] Nagano, Y. & Kim, C., A Two-Equation Model for Heat Transport in Wall Turbulent Shear Flows. *ASME Journal of Heat Transfer*, **110**, pp. 583–589, 1988.

[4] Nagano, Y. & Tagawa, M., An Improved $k-\epsilon$ Model for Boundary Layer Flows. *ASME Journal of Fluids Engineering*, **112**, pp. 33–39, 1990.

[5] Patel, V., Rodi, W. & Scheurer, G., Evaluation of Turbulence Models for Near-Wall and Low-Reynolds Number Flows. *3rd Symposium on Turbulent Shear Flows*, University of California, 1981.

[6] Chien, K., Predictions of Channel and Bounday-Layer Flows with a Low-Reynolds-Number Turbulence Model. *AIAA Journal*, **20**, pp. 33–38, 1982.

[7] Fan, S. & Lakshminarayana, B., Low-Reynolds-Number $k-\epsilon$ Model for Unsteady Turbulent Boundary-Layer Flows. *AIAA Journal*, **31**, pp. 1777–1784, 1993.

[8] Launder, J. & Sharma, *A first Course in Turbulence*. MIT Press: Boston, 1974.

[9] Škerget, L. & Ravnik, J., BEM simulation of compressible fluid flow in an enclosure induced by thermoacoustic waves. *Eng Anal Bound Elem*, **33**, pp. 561–571, 2009.

[10] Škerget, L., Hriberšek, M. & Žunič, Z., Natural convection flows in complex cavities by BEM. *Int J Num Meth Heat & Fluid Fl*, **13**, pp. 720–735, 2003.

[11] Popov, V., Power, H. & Škerget, L., (eds.) *Domain Decomposition Techniques for Boundary Elements: Applications to fluid flow*. WIT press, 2007.

[12] Ravnik, J., Škerget, L. & Hriberšek, M., 2D velocity vorticity based LES for the solution of natural convection in a differentially heated enclosure by wavelet transform based BEM and FEM. *Eng Anal Bound Elem*, **30**, pp. 671–686, 2006.

[13] Moin, P. & Kim, J., Numerical investigation of turbulent channel flow. *J Fluid Mech*, **118**, pp. 341–377, 1982.
[14] Kim, J., Moin, P. & Moser, R., Turbulence statistics in fully developed channel flow at low Reynolds number. *J Fluid Mech*, **177**, pp. 133–166, 1987.
[15] Kim, S., Near-Wall Turbulence Model and its Application to Fully Developed Turbulent Channel and Pipe Flows. *Numerical Heat Transfer,Part B*, **17**, pp. 101–122, 1990.
[16] Ramšak, M. & Škerget, L., A multidomain boundary element method for two equation turbulence models. *Eng Anal Bound Elem*, **29**, pp. 1086–1103, 2005.

# Numerical criteria for calculating the density diffusion in a water reservoir

K. Sakamoto[1], M. Kanoh[2] & T. Kuroki[3]
[1]*Environment Division, Matsue Doken Co. Ltd., Japan*
[2]*Department of Civil Engineering, Kyushu Sangyo University, Japan*
[3]*Baikoen 1-15-21 Dazaifu, Fukuoka, Japan*

## Abstract

In an earlier study, an attempt was made to ameliorate the concentration of oxygen in the lower layer of a reservoir by using a machine that supplies dissolved oxygen (DO). Field studies in a few water reservoirs have led to reports of a phenomenon in which the distance reached by the DO-rich water was more than 300 m (metres) in spite of the very low velocity of the water flow. In order to represent this phenomenon numerically, we proposed a velocity increase caused by the liquid density ρ, the gravity acceleration g and the time increment Δt. In this paper, we refer to call the velocity increase as the density diffusion, since the velocity increase seems to allow the area of DO diffusion to increase in the vertical direction. We would like to investigate the numerical criteria for calculating the density diffusion in a water reservoir using two-dimensional convective diffusion equations. Using the signs of the space division h (=Δs), the time increment k (=Δt), the diffusion parameter Γ (= $D*k/(h)^2$ ), and the Courant number $C_r$ (=V*k/h), we discuss the order estimate for calculating the density diffusion.

*Keywords: numerical criteria, density diffusion, meshless method, boundary element method, concentration in water reservoirs, observed concentration distribution in model simulation of water reservoir.*

## 1 Introduction

When the DO concentration equals 100 mg/L (milligrams per litre) the liquid density ρ becomes 1.0001 Kg/L (kilograms per litre). To estimate the liquid density ρ of 1.0001 Kg/L in the numerical analysis, it is necessary to ensure that

WIT Transactions on Modelling and Simulation, Vol 50, © 2010 WIT Press
www.witpress.com, ISSN 1743-355X (on-line)
doi:10.2495/BE100041

the space division h is less than $(0.0001)^{1/3}$ (=0.046) m in the calculation by using the method of the order of three-degree accuracy. If the order of two-degree accuracy is used, the space division h becomes less than $(0.0001)^{1/2}$ (=0.01) m. When diffusion parameter Γ, which is less than 0.288 in the calculation using the finite difference method (FDM), satisfies the order of four-degree accuracy ($O(h^4)$), the FDM can yield very accurate and convergent solutions when analysing the two-dimensional diffusion equation. To calculate the convection term, we use the FDM called the UTOPIA scheme or the QUICK scheme, which satisfies the order of three-degree accuracy ($O(h^3)$) (Leonard [2]). With respect to analysing the DO concentration by using the two-dimensional convective diffusion equation in the problem described above, we combine the FDM of the order of four-degree accuracy ($O(h^4)$) and the UTOPIA scheme or the QUICK scheme of the order of three-degree accuracy ($O(h^3)$). When the combined method above is adopted, the space division h should be less than $0.0001^{1/3}$ (=0.046) m for the estimation of the liquid density ρ of 1.0001 Kg/L. Next, we tried to upgrade the order of the accuracy of the meshless method, the BEM, and the finite element method (FEM). The upgrade was performed by introducing the radial basis functions of the Gaussians or the multiquadric to the meshless method, the special fundamental solution to the BEM, and the upwind shape function to the FEM. The newly developed methods, the meshless methods, the BEM, and the FEM, were tested to analyse the problem described above, and the order of the accuracy of these methods was analysed numerically. The calculated solutions obtained by using these methods were compared with the observed results in our model simulation, and the effectiveness and accuracy of the alternative numerical methods were estimated.

## 2 Governing equations

Convective-diffusion Equation (1) governs the diffusion of the concentration of oxygen in a water reservoir in the vertical ($x_1$, $x_2$) plane, as illustrated in Fig. 1,

$$C_{,t} + u_1 \cdot C_{,1} + u'_2 \cdot C_{,2} - D_1 \cdot C_{,11} - D_2 \cdot C_{,22} = 0 \qquad (1)$$

where C is the concentration of dissolved oxygen (DO), $C_{,t}$ is the time derivative of C, $u_1$ and $u'_2$ are the velocities of the $x_1$ and $x_2$ directions, respectively, and $D_1$ and $D_2$ are the diffusion coefficients of the $x_1$ and $x_2$ directions, respectively. Here, $C_{,1}$ and $C_{,2}$ describe the derivatives of C differentiated with respect to $x_1$ and $x_2$, respectively, and $C_{,11}$ and $C_{,22}$ are the derivatives of C differentiated twice with respect to $x_1$ and $x_2$, respectively. The velocity $u'_2$, which is shown in the above Equation (1), is defined as written in Equation (2),

$${}_{t+\Delta t}u'_2 = {}_t u_2 + (1 - \rho_{DO} / \rho_0)\rho g \Delta t = {}_t u_2 + \omega g \Delta t \quad \text{in the finite difference scheme} \qquad (2)$$

where ${}_{t+\Delta t}u'_2$ and ${}_{t}u_2$ are the velocities at time (t+Δt) and time (t) in the vertical direction, respectively. The second term (ωgΔt) of the right-hand side of Equation (2) means that the DO concentration increases the velocity of the vertical direction, and ω describes the density of the liquid that dissolves DO. Here, the velocity increase is caused by the liquid density ρ, the gravity acceleration g, and the time increment Δt. We refer to the velocity increase as the density diffusion, since the velocity increase seems to allow the area of DO diffusion to increase in the vertical direction, as described above, and expect that the velocity increase in the convective diffusion can be used as a device or evidence to explain the phenomenon in which the distance reached by the DO-rich water was more than 300 metres in spite of the very low velocity of the water flow    Here, the density ρ is connected to the DO concentration C, as written in Equation (3), where $\rho_0$ and $\rho_{DO}$ describe the densities of pure water and dissolved oxygen, respectively.

$$\rho = \rho_0 + C\cdot 10^{-6}(1-\rho_{DO}/\rho_0) \tag{3}$$

## 3 Numerical methods for calculating the density diffusion

We applied the meshless method, the FDM, the FEM and the BEM to analyse the density diffusion in the unsteady state in a water area, as shown in Fig. 1.

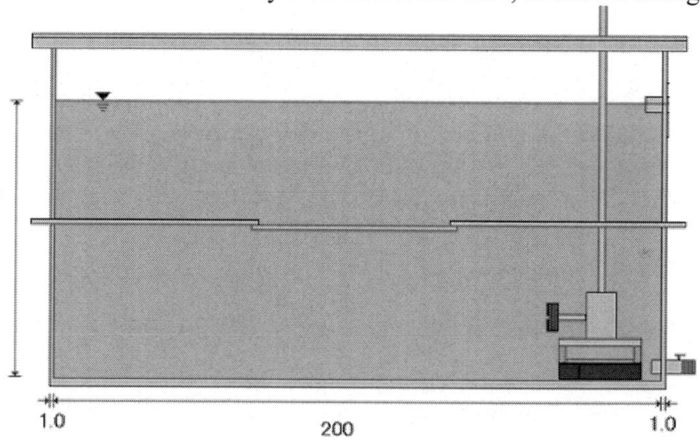


Figure 1: Analytical domain and a DO-supplying machine in a constructed model of a water reservoir.

### 3.1 Meshless method formulation for concentration analysis

The concentration in the steady state is expressed as Equation (4) with Equation (5) (Sakamoto et al. [1]),

$$C = \gamma_j Xj \ , \ Xj = (r^2+c^2)^{-1/2} \ \text{ or } \ Xj = \exp(-c\, r^2) \tag{4}$$

$$\left\{\left(\underline{u}_1 \frac{\partial X_j}{\partial x_1} + \underline{u}_2 \frac{\partial X_j}{\partial x_2}\right) - (D_1 \frac{\partial^2 X_j}{\partial {x_1}^2} + D_2 \frac{\partial^2 X_j}{\partial {x_2}^2})\right\}\gamma_j = 0 \tag{5}$$

where r equals $\{(x\text{-}x_j)+(y\text{-}y_j)\}^{1/2}$ and c is the constant. The transient convective-diffusion equation is then rewritten as follows

$$C_{,t} + L(C) = 0 \tag{6}$$

where $C_{,t}$ is the time derivative of C and L(C) has the terms of convection and diffusion in the steady state. Applying the finite difference scheme, Equation (6) yields

$$(C^{t+\Delta t} - C^t)/\Delta t + \{L^{t+\Delta t}(C) + L^t(C)\}/2 = 0 \tag{7}$$

$$C^{t+\Delta t} + L^{t+\Delta t}(C) \cdot \Delta t/2 = C^t - L^t(C) \cdot \Delta t/2 \tag{8}$$

where $C^{t+\Delta t}$ and $C^t$ are the concentrations at time (t+Δt) and time (t), respectively, and $L^{t+\Delta t}$ and $L^t$ are the terms of convection and diffusion at time (t+Δt) and time(t), respectively. Finally, using Equations (3), (4), and (8), the meshless method can be used to analyse the DO concentration in the unsteady state using the global expansion function Xj (= $(r^2+c^2)^{-1/2}$ or $\exp(-c\ r^2)$) of the mesh-free RBF collocation method (Divo et al. [3]) or the radial basis functions of the Gaussians (Powell [4]).

### 3.2 Finite difference method for convective-diffusion analysis

To analyse the DO concentration by using the two-dimensional convective - diffusion equation, we combine the finite difference scheme of the order of four-degree accuracy ($O(h^4)$) for the diffusion terms and the UTOPIA scheme of the order of three-degree accuracy ($O(h^3)$) for the convective terms (Leonard [2]). The weighted finite difference method (WFDM) (Kanoh et al. [5]) is also applied to convective-diffusion analysis.

### 3.3 Finite element method for convective-diffusion analysis

The upwind shape function (Kanoh and Kuroki [6]) and the ordinary shape function are tested to analyse the two-dimensional convective-diffusion problem using the FEM. The upwind finite element formulation is expected to yield a high order of accuracy to the computation of the problem, since the upwind weights of the exponential function give an exact solution to the one-dimensional convective -diffusion equation (Kanoh and Kuroki [6]). With respect to the analysis of the one-dimensional convective-diffusion equation, we propose that the analysis of the two-dimensional density diffusion be calculated by using the one-dimensional convective-diffusion analysis twice. The pure diffusion analysis can be perfectly calculated using the 7,500 elements (h=0.011m) with the ordinary shape function.

### 3.4 Boundary conditions and boundary discretisation

Both the boundary conditions and the boundary discretisation for the flow and concentration analyses have been previously proposed for the meshless method, the BEM, the FEM, the FDM, and the WFDM (Sakamoto et al. [1]).

## 4 Model simulation

We introduced a concept, described in the next subsection, in the simulation model constructed in our laboratory and obtained some observed velocity vectors and the distributions of the DO concentration in the model. In reference to the observed results, we tried to obtain some evidence to explain the phenomena that the distance reached by the DO-rich water was more than 300 metres in a reservoir in spite of the small velocity of the water flowing out. For this purpose, it was necessary to reproduce, in our model simulation, the density flow and convective diffusion of the DO concentration in the lower layer of a water reservoir at a depth of about 50 metres.

### 4.1 Simulation technique in the model

The concept introduced in our simulation model is described as follows: the density difference among 7 mg/L, 30 mg/L, and 100 mg/L in the DO concentration was changed to the density difference of the water temperature, since it was very difficult to make up the high concentration of DO of 100 mg/L in our model simulation at a depth of about 1.0 metres. Referring to Table 1, the density difference between 7 mg/L and 30 mg/L in DO was equal to the difference of the water temperature between 15.00°C and 14.87°C. Furthermore, the density difference between 7 mg/L and 100 mg/L in DO was equal to the difference of the water temperature between 15.00°C and 14.55°C. We iced the water that flowed out of the tank of the model and could easily control the difference of the water temperature among 15.00°C, 14.87°C, and 14.55°C. In this paper, the analogy between the differences in the water temperature and the DO concentration was proved by referring to the concentration distribution in the reservoir model visualised using a pigment and a VTR. The demonstration procedure is described as follows. First, we flowed out iced water at 14.87°C into water at 15.00°C with the same DO concentration. Secondly, we flowed out water with DO of 30 mg/L into water with DO of 7 mg/L with the same water temperature. Thirdly, we compared the VTR pictures of the movement of the area of the concentration distribution of both the differences of the water temperature and the differences of the DO concentration and observed that the VTR pictures of both movements were almost the same at every second. We consider that the analogy between the differences in the water temperature and the DO concentration can reproduce the horizontal direction of water flow and the convective diffusion of the DO of the water reservoir at a depth of about 50 metres in our model simulation. Here, 5 Kg/cm$^2$ (kilogram per square

Table 1: Density difference between 100 mg/L and 30 mg/L in DO and that of the water temperature among 15.00, 14.87, and 14.55 (°C).

| Place / Value | B water reservoir (waterdepth:50m) | Our model simulation (water depth:1m) |
|---|---|---|
| Maximum DO value (mg/L) | 100 | 30 |
| Water head (Kg/cm$^2$) | 5.0 | 0.1 |
| Value of ρ (using Equation (3): at 15°C) | 0.999127+0.0001 | 0.999127+0.00003 |
| Water temperature that corresponds to the above value of ρ (°C) | 14.55 (=15.0-0.45) | 14.87 (=15.0-0.13) |

centimetre) corresponds to 0.5 MPa (megapascal) in the international system of units (SI).

## 5 Results and discussion

As described above, we introduced a concept in the simulation model constructed in our laboratory and were able to observe some velocity vectors and obtain the distributions of the DO concentration in the model. In reference to the observed results, we tried to obtain some evidence to explain the phenomena that the distance reached by the DO-rich water was more than 300 metres in a reservoir in spite of the small velocity of the water flowing out. The numerical results of the meshless method, the BEM, the FEM, and the WFDM are also discussed in this section in order to investigate the numerical criteria for calculating the density diffusion in a water reservoir using two-dimensional convective diffusion equations.

### 5.1 Observed values in a model around a DO-supplying machine

#### 5.1.1 Flow velocity in a model

Figure 2(a) is an illustration of the velocity vectors of the temperature difference, -0.1, caused by a DO-supplying machine in a reservoir model visualised using a pigment (methylene blue), aluminium flakes, a strobe light, and a digital VTR. Figure 2(b) shows the velocity vectors of the DO concentration difference, 23 mg/L, in the reservoir model. We consider that the analogy between the differences of the water temperature and the DO concentration can be proved by using the observed velocity vectors, since both profiles of Fig. 2(a) and Fig. 2(b) seemed almost identical.

### 5.1.2 Concentration distribution of DO in a model

Figure 3 is an illustration of the concentration distribution of the temperature difference, -0.1, caused by a DO-supplying machine in a reservoir model visualised using a pigment (methylene blue) and a VTR. We consider that the analogy between the differences of the water temperature and the DO concentration can also be proved by using the observed concentration distribution, since both profiles of the temperature difference (Fig. 3) and the DO-concentration difference seemed almost identical (the figure to illustrate the concentration distribution of the DO-concentration difference, 23 mg/L, in the reservoir model was omitted).

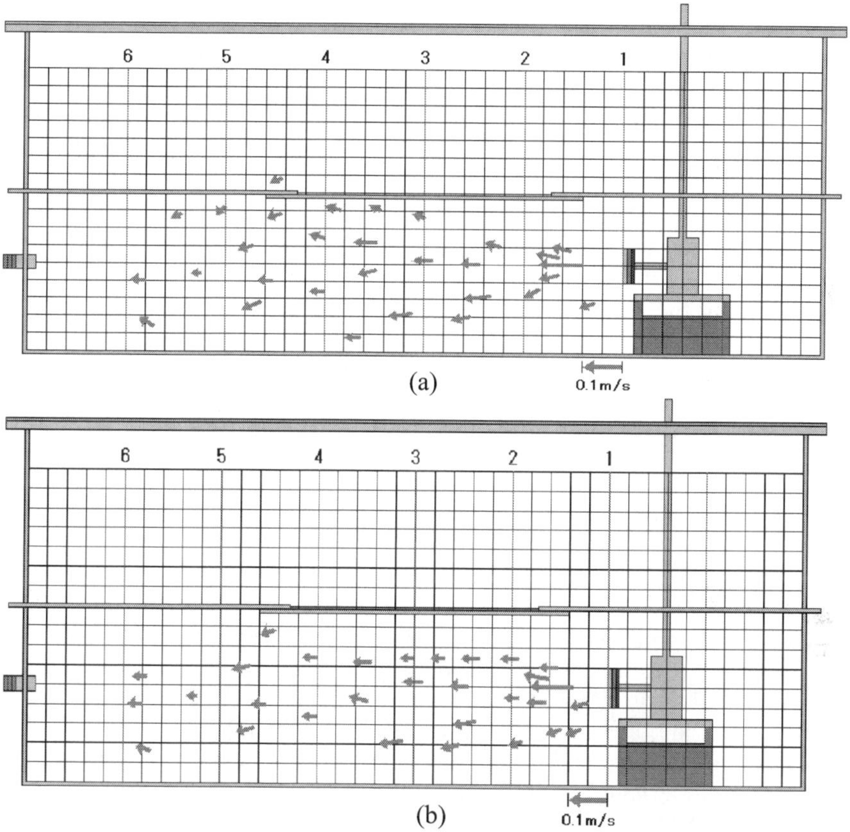


Figure 2: (a) Observed velocity (Temperature difference:-0.1). (b) Observed velocity vectors (DO difference: 23mg/L).

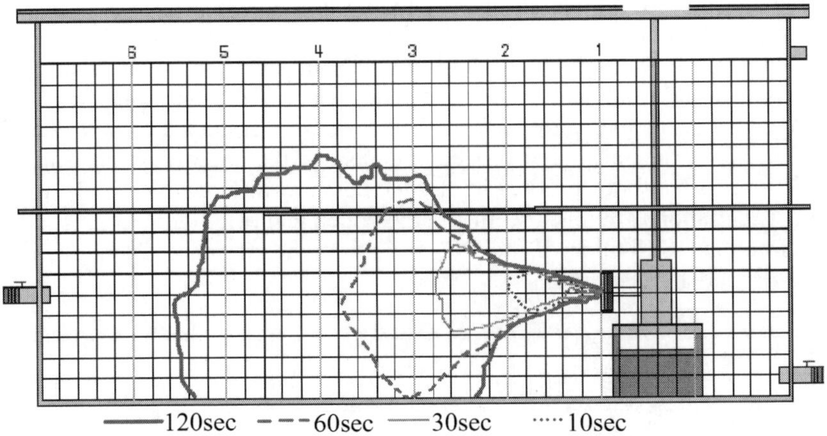


Figure 3: Observed areas of diffusion of pigment (temperature difference:-0.1).

### 5.2 Flow analysis in the model of a water reservoir

For computing the density diffusion, the velocity data are important and have significant influence on the calculated results. However, we would like to focus on the numerical criteria for calculating the density diffusion in this paper. The figures to illustrate the calculated velocity vectors were omitted, and discussions of flow analysis are limited to their influence for computing the density diffusion.

### 5.3 DO concentration analysis in the model of a water reservoir

#### 5.3.1 Time required by the four numerical methods for the DO analysis

Table 2 shows the time required by the four numerical methods for analysing the DO concentration in the model. When the number of divisions of the analytical domain was 4,961, the FDM, the FEM, and the meshless method needed almost 1.1, 4.4, and 10.3 times the time required by the WFDM, respectively. For the purposes of saving time, the WFDM was the best; the FDM was second best; the FEM was the third best; and the meshless method was the poorest performer. We believe that the reason that the WFDM was the best, the FDM was the second best, and the FEM was the third best is that the WFDM and FDM can be easily applied to an explicit scheme and the coefficient matrix of the FEM is suitable for employing the skyline solver. On the other hand, for the purpose of saving the time and labour required for preparing the input data, the meshless methodwas the best, the FDM was thesecond best, the FEM was the third best and the BEM was the worst.

#### 5.3.2 FEM calculation of the concentration distribution

Figures 4(a) and 4(b) are illustrations of the concentration distribution calculated using the FEM with the ordinary shape function, in which the number of

Table 2: The time required by the four methods for analysing the unsteady convective diffusion of DO for 240 seconds in a model of a water reservoir.

| Numerical method \ Relative computational time | Number of divisions: 4,961 | Time increment: Δt (sec) |
|---|---|---|
| WFDM | 1.00 | 0.005 |
| FDM | 1.10 | 0.005 |
| FEM | 4.40 | 0.1 |
| Meshless method | 10.3 | 5.0 |

divisions in the FEM is 4,800 and 10, 800, respectively. Here, the two values of $\lambda$ and $\nu$ are 1,000.0 and 0.0001 $m^2$/sec, respectively. Comparing Fig. 4(a) (in which the adopted number of divisions is 4,800) with Fig. 4(b) (in which the adopted number of divisions is 10, 800), it was noted that the increase of the number of divisions made the areas of the DO distribution wider in the vertical direction in the FEM analysis. With respect to the influence of flow analysis for computing the density diffusion, it is important to use an appropriate value of $\nu$ in the FEM flow analysis. The value of $\nu$ changed the velocity-vector

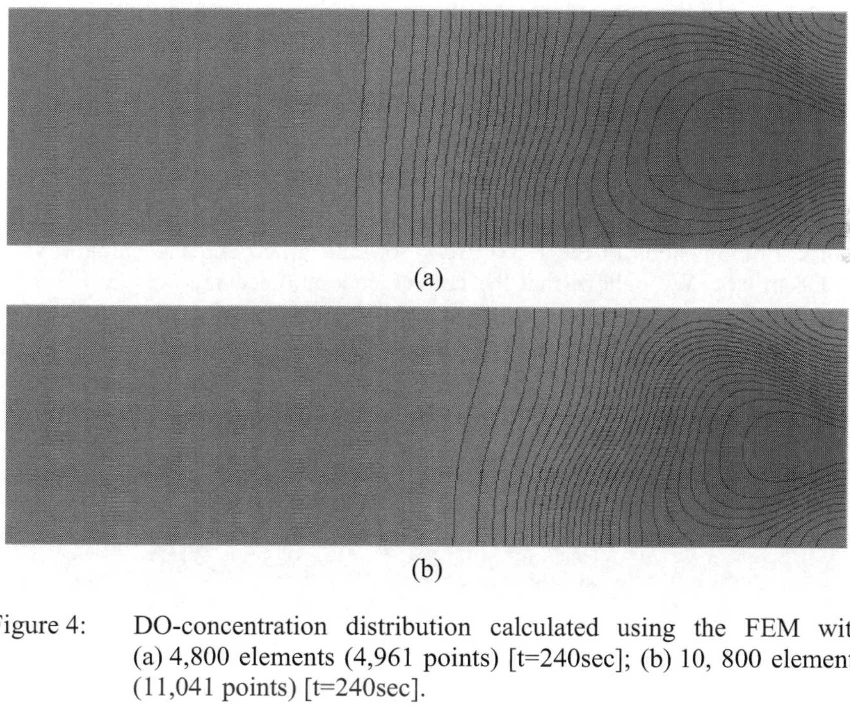

(a)

(b)

Figure 4: DO-concentration distribution calculated using the FEM with (a) 4,800 elements (4,961 points) [t=240sec]; (b) 10, 800 elements (11,041 points) [t=240sec].

distribution so significant that it was necessary to adopt the optimum value of ν. We believe that the upwind shape function may make it possible to set the optimum value of ν in the FEM flow analysis.

### 5.3.3 Meshless calculation of the concentration distribution

Figure 5 is an illustration of the concentration distribution calculated using the meshless method, in which the term of the velocity increase ($\omega g\Delta t$) is adopted, the number of divisions in the meshless method is 4,800, and the value of ν is 0.0001 $m^2$/sec. Referring to Figs. 4(a) and 5, the solutions of the meshless method showed the same tendency as those of the FEM in this problem. We considered that the convergence and accuracy of the FE, the FD, the BE, and the meshless methods for this problem were satisfactory (the figures to illustrate the concentration distribution calculated using the FEM with the upwind shape function and the BE method have been omitted).

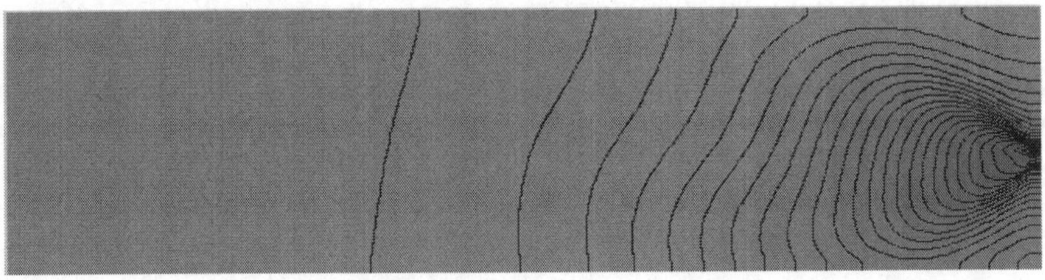

Figure 5: DO-concentration distribution using the meshless method with 4,961 points [t=240sec].

### 5.3.4 FDM calculation of the concentration distribution

Fig. 6 is an illustration of the concentration distribution calculated using the FDM, in which the finite difference scheme of the order of four-degree accuracy ($O(h^4)$) for the diffusion terms and the UTOPIA scheme of the order of three-degree accuracy ($O(h^3)$) for the convective terms are combined. Here, the numbers of divisions in the FDM are 4,800 and 8,800 and the value of ν is 0.0038 $m^2$/sec. We believe that the convergence and accuracy of the FDM for

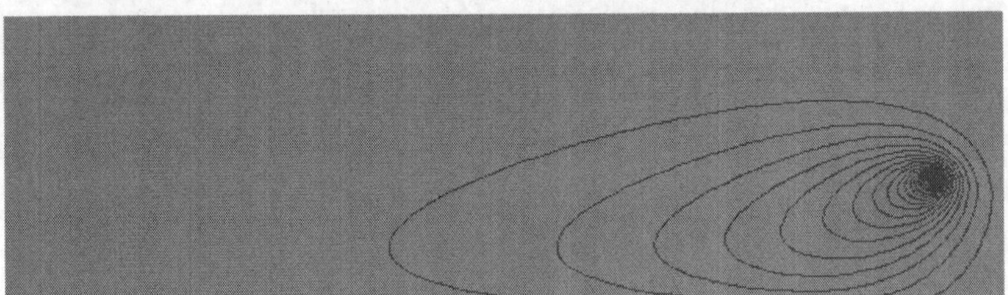

Figure 6: DO-concentration distribution using the FDM with 4,961 points [t=60sec].

this problem were satisfactory. The term (ωgΔt) of the density diffusion seemed to make the areas of the diffusion wider in the vertical and flowing-out directions and the speed of the convective diffusion higher than those in the analyses of the FDM when this density diffusion was not applied.

## 6 Conclusion

In summary, (1) in this paper, the velocity increase was defined as the density diffusion that was caused by the water density ρ, the gravity acceleration g, and the time increment Δt; (2) the meshless method, the BEM, the FEM, the FDM, and the WFDM were developed and applied to the analysis of the density diffusion; (3) introducing the radial basis functions of the Gaussians or the multiquadric to the meshless method, the special fundamental solution to the BEM and the upwind shape function to the FEM, we tried to upgrade the order of the accuracy these methods; (4) the finite difference scheme of the order of four-degree accuracy ($O(h^4)$) for the diffusion terms and the UTOPIA scheme of the order of three-degree accuracy ($O(h^3)$) for the convective terms were successfully combined for the FDM;(5) the density diffusion could make the areas of the diffusion wider in the vertical and outflow directions and make the speed of the convective diffusion be higher than in the analyses of these methods when this velocity increase was not applied; (6) the stability and convergence of the five kinds of analysis using these newly developed methods seemed satisfactory; (7) the analogy between the differences of the water temperature and the DO concentration could be proved by using both the observed concentration distribution and the visualised velocity vectors in our model simulation; (8) these developments and ideas described above were investigated, and the numerical criteria for calculating the density diffusion in a water reservoir using the two-dimensional convective diffusion equations was discussed.

## References


[1] Sakamoto, K., Kanoh, M., &., Kuroki T., Meshless and BE analysis of flow and concentration around a DO-supplying machine, *Proc. of the 31th World Conf. on Boundary Elements and Other Mesh Reduction Methods*, ed. C.A. Brebbia, WIT PRESS, Maribor, Slovenia, pp. 23-32, 2008.

[2] Leonard, B.P., The QUICK algorithm, Computer Methods in Fluids, ed. K. Morgan, C. Taylor, C.A. Brebbia, WIT PRESS, Pentech Press, p.159, 1980.

[3] Divo E., Kassab A., and Zahab El., Parallel domain decomposition meshless modeling of dilute convection-diffusion of species, *Proc. of 27th World Conf. on Boundary Elements and Other Mesh Reduction Methods*, ed. C.A. Brebbia, WIT PRESS, Florida, USA, pp.79-89, 2005.

[4] Powell, M.J.D, The theory of radial basis function approximation in 1990, *Advances in Numerical Analysis*, Vol. II, ed. W. Light, Oxford Science Publications: Oxford, pp.105-210, 1992.
[5] Kanoh, M., Kuroki T., Fujino K., and Ueda, T., Weighted Finite Difference and Boundary Element Methods Applied to Groundwater Pollution Problems, *Wat. Sci. Tech. Vol. 23, Kyoto,* pp. 517-524, 1991.
[6] Kanoh, M. and Kuroki T., Upwind finite element analysis of the inverse problem for groundwater resources, *Proc. of 2nd International Symposium on Inverse Problems - ISIP '94, Paris,* ed. Bui, H.D., Tanaka M. et al., pp. 29-32, 1994.

# Optimum design of cut off walls located in the foundation of diversion dams using boundary element method

S. A. Lashteh Neshaei, M. A. Mehrdad & N. Abedi Mahzoon
*Department of Civil Engineering, University of Guilan, Rasht, Iran*

## Abstract

Foundation design for hydraulic structures founded on permeable soils has been a concern among designers. Estimation of uplift forces and hydraulic gradient at any point, especially at key locations of a permeable media is one of important elements in designing concrete gravity dams on such foundation conditions. Among different numerical methods, the Boundary Element Method (BEM) employed for solving Laplacian equation of flow through porous media has been proved to be a powerful and effective tool for design purpose. The precision and speed of this method to evaluate the hydraulic gradient and potential at any point of foundation enable designers to find a suitable location of cutoff walls and filters.

In this investigation, the impression of various parameters such as length and position of cutoff walls, position of filter, depth of porous foundation and type of soils has been considered and their effects on uplift pressure and hydraulic gradient are represented graphically. In particular, the results obtained from the present work clearly indicate that placing an extra cutoff wall between the two lateral walls has no considerable effect on the uplift force or hydraulic gradient, whereas locating a proper filter in a suitable place on the foundation can remarkably change the flow pattern and affect the design process. Results of this study could be of the great help for engineers to design diversion dams by efficient methods.

*Keywords: BEM, diversion dam, uplift pressure, hydraulic gradient, cut off wall, filter.*

WIT Transactions on Modelling and Simulation, Vol 50, © 2010 WIT Press
www.witpress.com, ISSN 1743-355X (on-line)
doi:10.2495/BE100051

# 1 Introduction

Installation of hydraulic structures in areas where water penetration under facilities is probable requires great care due to instabilities caused by the leakage of water. Mentioned instabilities occur mainly due to two reasons, the first is uplift pressure development and change in the equilibrium of effective forces and the second reason involves internal gradual erosion of foundation material resulting in the piping phenomenon. Therefore, in all cases which foundation of a structure constructed on permeable bed, it is necessary that bearing pressure on the contact surface between foundation and structure and also output hydraulic gradient in down-level of contact surface to be estimated. For example, such studies about diversion dams located on the bed of alluvial river is crucial, because of upstream water storage and then creating the headwater differential in upstream and downstream of the dam, seepage under the foundation will be occurred and this may cause pressure to the contact surface between foundation and dam.

Different methods have been applied to make economic plans to reduce uplift pressure in the concrete gravity dams, some of the most important of them can be mentioned as the bed filter and cut off walls which are placed on the upstream and downstream of the dam (to increase the flow path length and creating more head loss) and construction of drain on the downstream of the dam or in the appropriate place between the two cut off walls (to reduce output gradient) [2]. Mainly, the present study is aimed to investigate the effects of these parameters to reduce output gradient and uplift pressure. A mathematical model is presented to show the effects of number, location and length of cut off walls and drain on uplift pressure and output hydraulic gradient at key locations under diversion dams and general stability of dam has been evaluated using Mseep program. The fundamental concepts of Mseep program are based on the Finite Element Method while the proposed model applied in this study utilizes the Boundary Element Method.

# 2 Theoretical development

The Laplace equation is the governing equation for two-dimensional potential flows in porous medium which reads as:

$$\nabla^2 \phi = \frac{\partial^2 \phi}{\partial X^2} + \frac{\partial^2 \phi}{\partial Y^2} = 0 \tag{1}$$

$\phi$ in this equation is the potential function and flow velocity components will be derived from this function. Solution of Laplace's equation and determination of the potential function at key locations will result in the specifications of flow. To solve the Laplace's equation different methods have been suggested among them the most important are:

1. Laboratory methods constructed based on physical models.
2. Graphical methods in which the $\phi$ values obtained by drawing the flow lines and equipotential lines (flow net).

3. Theoretical - empirical methods (such as the Lane and Bligh methods).
4. Combined potential method which based on complex functions leed to analytical solution of equations.
5. Numerical methods such as the Finite differences, Finite elements, and Boundary Elements.

Nowadays, numerical methods for solving Laplace equation are used widely and have become common. High speed in determination of the potential and hydraulic gradient values in all parts of a permeable field is the advantage of these methods. One of the most appropriate methods in solving the Laplace's equation is the Boundary Element Method (BEM), in which the approximate shape functions are used so that within the specific domain satisfied governing differential equation on the field but not at the boundaries [4]. Unknown coefficients of these functions are calculated through applied boundary conditions at certain points at the boundaries. Therefore unknown function values at the boundaries are obtained completely and then with the position of internal domain points, unknown function values of the points can be calculated. A sample of such networking system is shown in fig. 1. Concept of these elements is that the potential function is constant along the elements and equal to the potential value at the center of element. This assumption could not be very unreasonable, if large numbers of elements are selected.

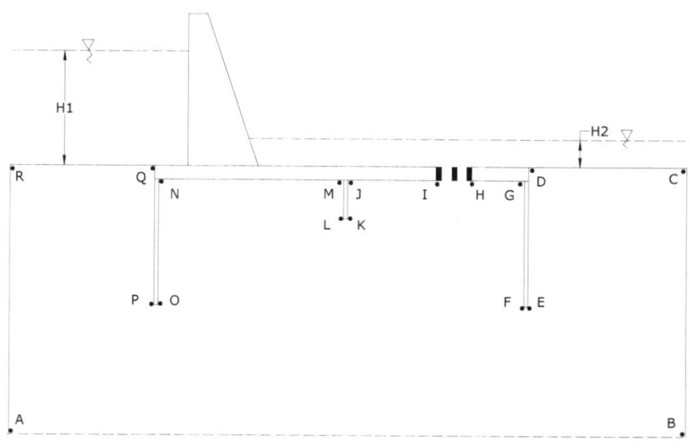


Figure 1: Schematic networking of mathematical model of the foundation of concrete gravity dam.

The governing equation in BEM can be written in a matrix form as:

$$[H]\{\emptyset\} = [G]\{q\} \tag{2}$$

In which:

$\{\emptyset\}$: Potential vector in elements.

$\{q\}$: Velocity vector in elements.

$[H]$ and $[G]$: matrix of coordinates which defined as:

$$H = \int_\eta \frac{\partial \emptyset^*}{\partial r} d\eta \tag{3}$$

$$G = \int_\eta \emptyset^* \, d\eta \tag{4}$$

$$\emptyset^* = \frac{1}{4\pi r} \tag{5}$$

in which:
r: distance from the centre of element.

Solving the above equations will lead us to the magnitudes of $\emptyset$ in the nodes; and and subsequently the pressure and hydraulic gradient can be derived from the potential function.

Variables in the present model which illustrated in fig. 2 are as follows [3]:

H1 = water head on upstream.
H2 = water head on downstream.
$\Delta$H = height of effective water.
T = depth of permeable layer.
$D_1$ = length of upstream cut off wall.
$D_2$ = length of downstream cut off wall.
L = length of bed filter.
$L_1$ = distance from the middle cut off wall to the upstream cut off wall.
$L_2$ = distance from the drain to the upstream cut off wall.
f = length of drain.
t = thickness of the cut off wall.
t'= thickness of bed filter.

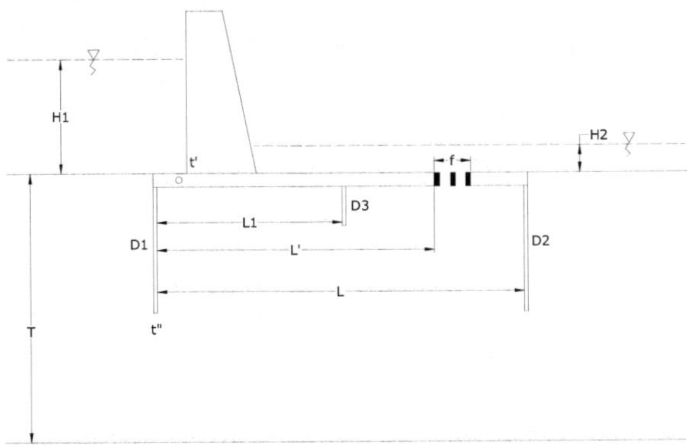


Figure 2: Variables in mathematical model of the foundation of concrete gravity dam.

The reasons for applying the boundary elements method to this particular case can be explained as follows:

1. Laplace equation can be easily solved by this method and the results are satisfactory.

2. Due to less degree of freedom in this method rather than other numerical methods (such as the Finite Element Method), the number of equations to be solved in the present method under the similar conditions will be less.
3. In the present method, networking of system is easily possible due to one-dimensional nature of elements.
4. Major changes in the system, in this particular case, generally are at the boundaries, thus using the Boundary Element Method (BEM) which analyzed networked boundaries will be simpler than other numerical methods.

Boundary in the present system, including boundaries A, B, C, D, E, F, G, H, I, J, K, L, M, N, O, P, Q, R can be divided into two parts, fig. 1, [1]:

1. Boundaries that have the boundary conditions of the type of $\phi = \overline{\phi}$. These borders include some points in which the potentials are known (e.g. RQ, IH and DC boundaries in the fig. 1 that amount of piezometric heads are equal to values H1, H2 and H3)respectively. It is to be mentioned that if the Laplace equation needs to be considered in the form of $\nabla^2 h = 0$ (h is the piezometric head), system boundary conditions will be of the type of $h = \overline{h}$ .

2. Boundaries that have the boundary conditions of the type of $\frac{\partial \phi}{\partial n} = \overline{q}$ : in this boundaries flow velocity (derivative of the potential function) is a certain amount (n is unit vector perpendicular to the surface). For example, boundaries of impermeable layer, cut off walls and bed filter in fig. 1, include this type of boundaries (e.g. AB, BC, DE, EF, FG, and GH boundaries and other similar boundaries).

After introduction of the boundaries and sorting system network into boundary elements with certain geometric model and node coordinates, the computer can easily solve boundary element method equations system and find the potential and hydraulic gradient values at specific points. Noting the following comments, [5], in the model networking are necessary:

1. System lateral boundaries should be considered so that the effect of the uplift pressure at those is negligible.
2. Flow velocity at the lateral boundaries is considered to be equal to zero (similar to impermeable layer line AB).
3. At the cut off wall, bed filter and impermeable layers, flow velocity is equal to zero.
4. On the levels of the ground under the bed filter, water head has specific values and is equal to the height of water above lines.
5- Along the drain the piezometric head like the uplift pressure is equal to H2.
6- Elements numbering is conducted in an arbitrary direction (for instance in counter clockwise) in accordance with the order.
7. Capability of the present model is in homogeneous and isotropic soils and in other situations equivalent permeability coefficient should be used.
8. One of the advantages of this model is that, any change in the system boundaries can be reviewed with the introduction of boundary conditions related to it.

9. To evaluate the effect of each variable, other parameters are assumed to be constant, and eventually led to the charts that show the effect of proposed variable on the uplift pressure or output hydraulic gradient.
10. Due to the importance of uplift pressure at point G and the hydraulic gradient at point D, fig. 1, only these two key locations are taken into account.

Using Mseep program, parameters in fig. 2 have been taken into the program as input and ultimately, pressure and potential values have been calculated at any location on the permeable layer under the dam. Then the results have been plotted in some diagrams. The following limitation exists in the modelling procedure:
1. To consider the effect of various parameters, other parameters are assumed to be constant.
2. Due to the low importance of thickness of the cut off wall, it is assumed to be constant.
3. Instead of water heads at upstream and downstream to be considered as two different parameters, the difference of the water head between upstream and downstream (H) is considered as a critical parameter.
4. Maximum hydraulic gradient considered at the end of the dam impermeable bed, because of the risk of piping in that region.
5. Because of the existence of filter, the variation of the uplift pressure is not uniform and in this study the average uplift pressure value is calculated and analyzed.

## 3 Analyses of the results

Analyses of the uplift pressure and the hydraulic gradient output values calculated at key locations on boundaries with the Mseep program and present methods lead to different designed curves. The following results can be obtained from this investigation:
1. Increasing the length of upstream cut off wall (D1) results in decreasing the uplift pressure value at point G and reducing the output hydraulic gradient at point D caused by increased in the flow length along the way. Increasing distance from the drain to the cut off wall causes the output gradient value to rise at point D, but uplift pressure might increase or decrease at point G (most of the curves have the uplift pressure reduction). Increasing the output gradient due to more density of input flow lines on the foundation can be expected when the drain moves to downstream, and due to low density of the flow lines after drain, the uplift pressure will be decreased at point G. The results are consistent with the results of the Mseep program, fig. 3.
2. Placing the filter upstream results in the uplift pressure and hydraulic gradient values to be decreased, but at the vanes of end cut off wall, the uplift pressure as a factor of instability will be increased [6]. Fig. 4 shows the results of Mseep program which confirm the results of the present investigation.

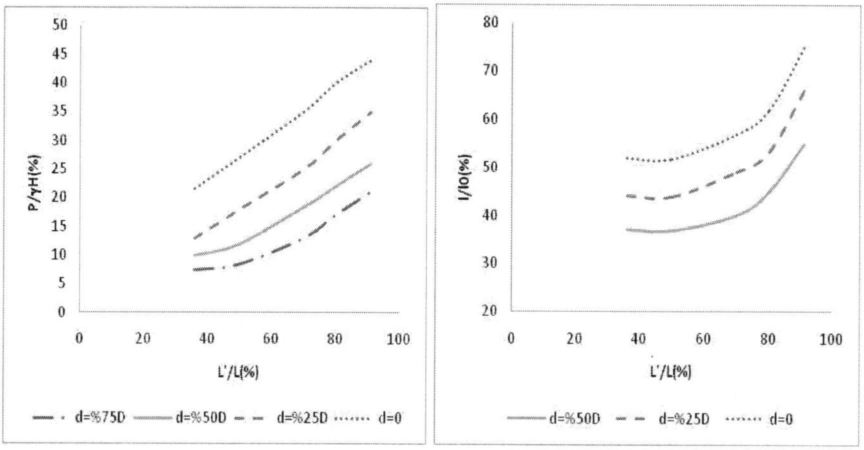


Figure 3: Combined effect of locating filter and length of the primary cut off wall on uplift pressure and hydraulic gradient.

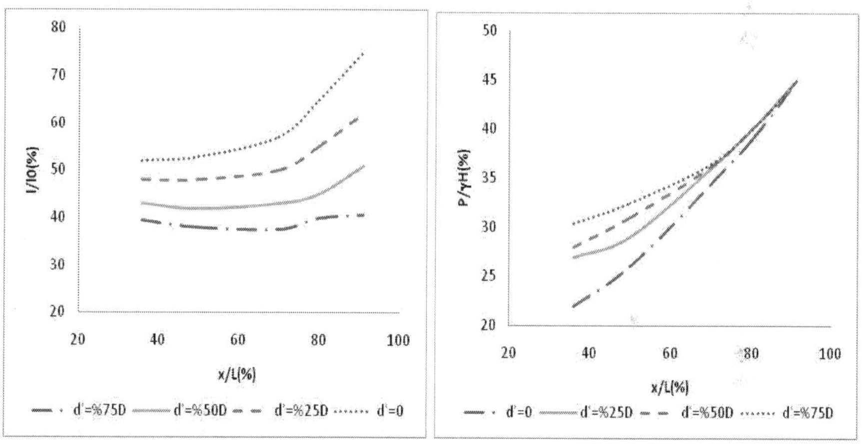


Figure 4: Combined effects of the location of filter and the length of the lateral cut off wall on uplift pressure and hydraulic gradient.

3. Increasing the drain length increase (f) makes a significant decrease in uplift pressure and especially in the output hydraulic gradient, fig. 5.
4. Probability of the piping phenomenon (hydraulic gradient reaches to a critical value) is possible in high effective head (about four times depth of permeable foundation). The occurrence of such a potential is unlikely in a small diversion dam [7]. In other word, it can be concluded that in small diversion dams, the hydraulic gradient cannot be an effective design parameter because in the normal cases it never reaches to the critical limit. Fig. 6 shows the results of the Mseep program which confirms the mentioned result.

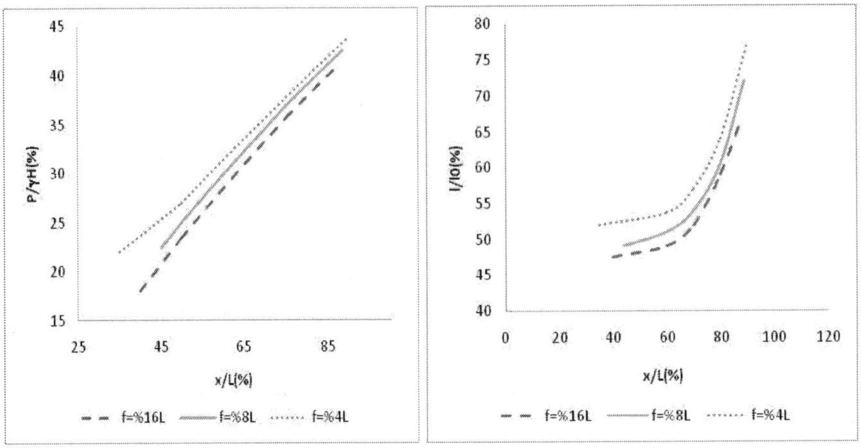


Figure 5: Effect of filter width on the uplift pressure and hydraulic gradient.

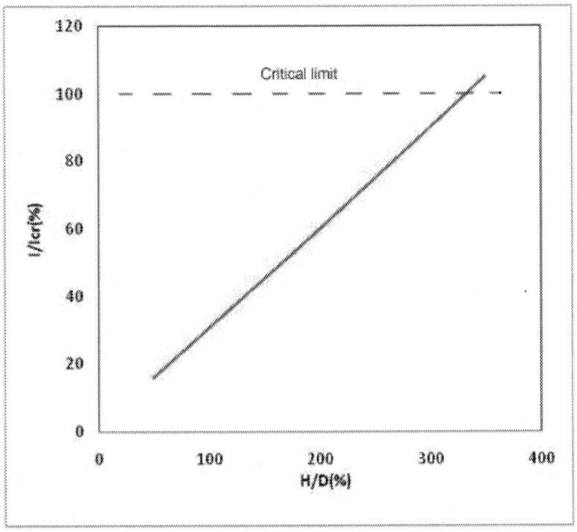


Figure 6: Probability of the piping phenomenon in small concrete diversion dam with a filter.

## 4 Summary and conclusions

The results obtained from the present model and comparison with those obtained from Mseep program indicates the following conclusions:

1. The cut off wall located upstream reduces the uplift pressure and output hydraulic gradient; both of them are desirable for designers and suitable for stability of the dam.

2. The cut off wall located downstream reduces the output hydraulic gradient; but it increases the uplift pressure which is considered as a negative factor on stability.
3. Filter (drain), reduces the uplift pressure and output hydraulic gradient. Reducing pressure is considerably more where the filter is located upstream of dam.
4. In small diversion dams the hydraulic gradient is less than its critical value and it is not considered as a major parameter in design.
5. Filter, itself, does all the effective functions of cut off wall without the negative effects. This effect will increase where filter is close to the upstream of dam and in this case, filter does the functions of the end cut off wall, and if the output discharge does not create any limit for designer, the end cut off wall can be shortened or even eliminated.

By comparing the results of the present model and Mseep program, it can be concluded that the present method introduces faster and easier approach to determine the output hydraulic gradient and uplift pressure with simple assumptions and solving few equations. An economical plan could be achieved for the optimized design of the foundation of diversion dams that minimizes the total cost of project with the results obtained from the proposed mathematical model. Dimensions of the cut off wall and drain will be designed to minimize the output hydraulic gradient and uplift pressure of dam foundation, with reduced concrete to the lowest volume as much as it is possible.

## References


[1] Berbbia, C., computational hydraulics, Butterworth publishing, London, 1983.

[2] Braja, M.D., Advanced soil mechanic, Mac Graw Hill Bok Company, New York, 1987.

[3] Chawla, A.s. & Kumar, A., Design of structures with Intermediate Filters, 1986, ASCE, Journal of Hydraulic Engineering, Vol. 112, No.4, pp. 835-847.

[4] Chung A.K. & Falvey, H.T., Seepage Analysis Using the Boundary Element Method, United state Department of the Interior, Bureau of Reclamation, 1984.

[5] Deshmukh, A.M., Khanolkar, B.V. & Dharward, R.C., Plastic cut off wall for seepage control at Salauli Dam, 8th Regional Conference on Soil Mechanics and Foundation Engineering, Kyoto, pp. 449-452, 1987.

[6] Liu, H. & Song, E., Working mechanism of cut off walls in reducing uplift of large underground structures induced by soil liquefaction, Computers and Geotechniques, Vol. 33, pp. 209-221, 2006.

[7] Uromeihyand, A. & Barzegari, G.,Evaluation and treatment of seepage problems at Chapar-Abad Dam, Engineering Geology, Iran, Vol. 91, pp. 219-228, 2007.

# Section 2
# Damage mechanics and fracture

# Integral equations for crack systems in a slightly heterogeneous elastic medium

A. N. Galybin[1] & S. M. Aizikovich[2]
[1]*Wessex Institute of Technology, Southampton, UK*
[2]*Rostov-on-Don Technical University, Russia*

## Abstract

The paper deals with derivation of a system of singular integral equations for slightly heterogeneous media. The system is derived in terms of complex potentials by introducing a small parameter by the perturbation method. The resulting system of integral equations is of the Cauchy type with the correction terms that address inhomogeneity of the media, they are presented in the form somewhat similar to the problem with body forces. Therefore the numerical methods developed for the homogeneous crack problems can directly be applied for the further numerical analysis.

*Keywords: heterogeneous media, plane elasticity, singular integral equations.*

## 1 Introduction

Elastic properties of many natural and artificial materials have slight fluctuations. They can be produced during formation of a material (e.g., metal sheets or concrete consolidation), loading history (rocks) or non-homogeneity of structure (ceramics). They can also be imposed for the sake to improve performance of particular materials as in fractionally graded materials, FGM. Non-homogeneity is often neglected for the determination of the fracture resistance of such materials. However, in some cases the effect induced by moduli fluctuations can create significant fluctuations of material fracture toughness, which is the case for FGM.

It should be noted that by introducing non-homogeneity of the fracture toughness one can model influence of stress fluctuations. However this case is fully not the fluctuations of material characteristics. They cannot be considered to be independent of applied load and therefore the direct superposition of two

WIT Transactions on Modelling and Simulation, Vol 50, © 2010 WIT Press
www.witpress.com, ISSN 1743-355X (on-line)
doi:10.2495/BE100061

solutions for homogeneous and stress fluctuations may produce significant errors. A number of papers have been published on the problems dealing with the crack propagation in heterogeneous materials, e.g. [1–7]. However the case of multiple cracks in heterogeneous media has not been addressed properly. The aim of this study is to derive integral equations describing equilibrium of a system of 2D cracks in a slightly heterogeneous media. The latter means that the fluctuations of the elastic moduli is much less that their average values. For simplicity, the consideration is restricted to the case of a thin plate with the crack, which makes it possible to use the methods of complex variables.

The approach is based on the use of complex potentials and the perturbation technique. First a general approximate solution for the plane problem is obtained by the method developed by Vekua [8, 9]. The form of this solution is somewhat similar to the solution presented in [10] for the case of body forces but contains two additional terms reflecting inhomogeneity. Then the system of equation is derived by the methods based on complex potentials [10] for an arbitrary smooth contour. This is a generalisation of the approach used in [7] for a single strait crack.

## 2 General solution for slightly inhomogeneous media by the perturbation method

### 2.1 Equations of plane elasticity for inhomogeneous medium in complex coordinates

The complete system of governing equations of 2D theory of elasticity consists of differential equilibrium equations (DEE), Hooke's law, strain-displacement relationships (SDR) and compatibility (of deformation) equation. By introducing complex coordinates $z = x + iy$ and $\bar{z} = x - iy$ and the following differential operators

$$\begin{aligned} 2\partial_z f &= \partial_x f - i\partial_y f, \quad 2\partial_{\bar{z}} f = \partial_x f + i\partial_y f \\ \partial_x f &= \partial_z f + \partial_{\bar{z}} f, \quad \partial_y f = i(\partial_z f - \partial_{\bar{z}} f) \end{aligned} \tag{1}$$

one can present the governing equations as follows.
The DEE assume the form

$$\partial_{\bar{z}} D = \partial_z P \tag{2}$$

where the stress functions $P$ and $D$ are used *instead* of the stress components $\sigma_{xx}$, $\sigma_{xy}$, $\sigma_{yy}$ in Cartesian coordinates

$$P = \frac{1}{2}(\sigma_{xx} + \sigma_{yy}), \quad D = \frac{1}{2}(\sigma_{yy} - \sigma_{xx}) + i\sigma_{xy} \tag{3}$$

Similarly to the stress function in (3) one can introduce the following strain function

$$\varepsilon = \frac{1}{2}(\varepsilon_{xx} + \varepsilon_{yy}) \quad \delta = \frac{1}{2}(\varepsilon_{yy} - \varepsilon_{xx}) + i\varepsilon_{xy} \tag{4}$$

and the complex displacement vectors

$$w = u_x + iu_y \tag{5}$$

where $u_x$ and $u_y$ are the displacement components and $\varepsilon_{xx}$, $\varepsilon_{xy}$, $\varepsilon_{yy}$ are strain components in Cartesian coordinates. Then the SDR assume the form

$$2\varepsilon = \partial_z w + \partial_{\bar{z}} \overline{w}, \quad \delta = -\partial_z \overline{w} \tag{6}$$

In 2D the equation expressing compatibility of strain components has the form

$$2\partial^2_{z\bar{z}}\varepsilon + \partial^2_{zz}\overline{\delta} + \partial^2_{\bar{z}\bar{z}}\delta = 0 \tag{7}$$

For an isotropic material the Hooke's law takes the form

$$P = \mathrm{K}\varepsilon, \quad D = 2\mu\delta \tag{8}$$

where μ is the shear modulus and K is a bulk modulus; for heterogeneous media they are spatial functions of coordinates.

The Lame equations (in terms of displacements) can be obtained by substitution of (8) into (2) followed by replacement of the strain functions via the SDR in the form (6). This gives

$$\partial_z\left(\mathrm{K}(\partial_z w + \partial_{\bar{z}}\overline{w})\right) + 4\partial_{\bar{z}}\left(\mu\partial_z\overline{w}\right) = 0 \tag{9}$$

If the problem is formulated in terms of stresses alone, then the complete system consists of the DEE (2) and the compatibility equation that is obtained by replacing the stain functions in (7) via the stress functions via the Hooke's law (8). In this case (7) assumes the form

$$4\partial^2_{z\bar{z}}\frac{P}{\mathrm{K}} + \partial^2_{zz}\frac{\overline{D}}{\mu} + \partial^2_{\bar{z}\bar{z}}\frac{D}{\mu} = 0 \tag{10}$$

It is also possible to express the stress functions via the Airy's function and substitute it to (7), which results in a single governing equation.

It is evident that for homogeneous media ($\mathrm{K}=\mathrm{K}_0$, $\mu=\mu_0$) the Lame equations take the form

$$\partial^2_{zz} w + \kappa_0 \partial^2_{\bar{z}z}\overline{w} = 0 \tag{11}$$

where $\kappa_0=3-4\nu_0$ for plane strain and $\kappa_0=(3-\nu_0)(1+\nu_0)^{-1}$ for plane stress and $\nu_0$ is homogeneous Poisson's ratio.

At the same time it is evident that both functions $P$ and ε are harmonic, i.e they satisfy he Laplace equation

$$\partial^2_{z\bar{z}}\varepsilon = 0 \quad \partial^2_{z\bar{z}}P = 0 \tag{12}$$

while the functions $D$ and δ are bi-holomorphic, i.e. their second derivatives with respect to the conjugated variable vanish

$$\partial^2_{\bar{z}\bar{z}}\delta = 0, \quad \partial^2_{\bar{z}\bar{z}}D = 0 \tag{13}$$

For heterogeneous media relationships (12) and (13) are not valid.

 
www.witpress.com, ISSN 1743-355X (on-line)

## 2.2 Approximation for a medium with fluctuating Young's modulus

We further consider the case when the Poisson's ratio is constant, $\nu_0$, therefore the following ratio is also constant

$$\frac{4\mu}{K} = \kappa_0 - 1 \tag{14}$$

Therefore the Young modulus, $E$, is the only deformation characteristic that varies in space. It is assumed that these variations are much less than the average value of the Young modulus, $E_0 = \langle E \rangle$, over the domain considered.

Let us present the Young modulus as follows

$$E = E_0 + \Delta E \tag{15}$$

Here the variations are designated as $\Delta E$ which is a known continuous function with respect to both variables and possessing the following property

$$\|\Delta E\| << E_0 \tag{16}$$

where $\|...\|$ stands for a norm; in particular, the norm can be defined as a global maximum of the function over the entire domain. Inequality (12) allows one to introduce a small parameter $\lambda$ as follows

$$\lambda = \max|\Delta E| E_0^{-1} \tag{17}$$

Hence (10) can be rewritten in the form

$$E = E_0(1 + \lambda e) \tag{18}$$

where $\|e\| \le 1$ and $0 < \lambda << 1$.

Substitution of (14) and (18) into (9) results in the following equation

$$\partial_{zz}^2 w + \kappa_0 \partial_{\bar{z}\bar{z}}^2 \overline{w} + \lambda\left[\partial_z\left(e(\partial_z w + \partial_{\bar{z}} \overline{w})\right) + (\kappa_0 - 1)\partial_{\bar{z}}\left(e\partial_z \overline{w}\right)\right] = 0 \tag{19}$$

Let us further seek solution for $w$ and for all other stress and strain functions in the form

$$Q = Q_0 + \lambda Q_1 + \lambda^2 Q_2 + \dots \tag{20}$$

where $Q$ can be $w$, $P$, $D$, $\varepsilon$, or $\delta$.

By substituting (20) into (19) one obtains the following recurrent process for solving the governing equation by successive approximations

$$\begin{aligned} &\partial_{zz}^2 w_0 + \kappa_0 \partial_{\bar{z}\bar{z}}^2 \overline{w}_0 = 0 \\ &\partial_{zz}^2 w_j + \kappa_0 \partial_{\bar{z}\bar{z}}^2 \overline{w}_j = -F_{j-1}, \quad j = 1,2,\dots \end{aligned} \tag{21}$$

where the right hand side in the second equations in (21) has the form

$$F_j = \partial_z\left(e(\partial_z w_j + \partial_{\bar{z}} \overline{w}_j)\right) + (\kappa_0 - 1)\partial_{\bar{z}}\left(e\partial_z \overline{w}_j\right) \tag{22}$$

It is known from the solution at the previous iteration.

It is evident that general solution from the zero iteration coincides with the general solution for homogeneous media presented by the Kolosov-Muskhelishvili formulae [10]

$$2\mu_0 w_0 = \kappa_0 \varphi_0 - z\overline{\varphi_0'} - \overline{\psi}_0, \quad P_0 = \varphi_0' + \overline{\varphi_0'}, \quad D_0 = \bar{z}\varphi_0'' + \psi_0' \tag{24}$$

Here $\varphi_0=\varphi_0(z)$, $\psi_0=\psi_0(z)$ are holomorphic functions inside the domain considered and the modulus $\kappa_0$ is given by (14).

### 2.2.1 First approximation

Let us find the first approximation for $Q$. First let us specified the right hand side of the second equation in (21) by substituting expression from (24)

$$F_0 = (\kappa_0 - 1)\left[\partial_z(eP_0) - \partial_{\bar{z}}(eD_0)\right] \tag{25}$$

Integration of the second equation in (21) can be obtained on the basis of the Pompeiu's formula [8], which has been derived for solving the non-homogeneous Cauchy- Riemann system. In the complex variables this system can be written by a single equation of the form

$$\partial_{\bar{z}} w = f \tag{26}$$

Solution of (24) is given by the following formula

$$w = \phi + Tf, \quad Tf = -\frac{1}{\pi}\iint\limits_{\Omega} \frac{f(\xi,\eta)}{\zeta - z} d\xi d\eta, \quad \zeta = \xi + i\eta, \quad z = x + iy \tag{27}$$

where $\phi=\phi(z)$ is an arbitrary holomorphic function in the considered domain $\Omega$. This representation contains the operator $Tf$ expressed by the double integral over the domain. However, formal integration of the second equation in (26) employs indefinite integrals with respect to independent variables $\bar{z}$ which gives

$$w = \phi + \int f(z,\bar{z}) d\bar{z} \tag{26}$$

Vekua [8] has demonstrated equivalence of both this presentations for analytical functions of two variables which can be expanded into power series with respect to these variables. This is a wide class of functions that covers many applications, therefore we further use indefinite integration. In the general case indefinite these integrals should be replaced by double integrals as in (25).

The general solution for $w_1$ can be presented as the sum of the general solution of the homogeneous equation, which has the form similar to that in (24) and a partial solution $W = W(z,\bar{z})$ of the non-homogeneous equation.

$$2\mu_0 w_1 = \kappa_0 \varphi_1 - z\overline{\varphi_1'} - \overline{\psi}_1 + W \tag{27}$$

where new holomorphic functions $\varphi_1=\varphi_1(z)$, $\psi_1=\psi_1(z)$ have been introduced

After substitution of (27) into the left hand side on the second the function $W$ can be expressed in the following form

$$(\kappa_0^2 - 1)W = \int \left(\kappa_0 \overline{G_0} - G_0\right) dz, \quad G_0 = \int F_0 dz \tag{28}$$

At the same time the stress functions presume the form

$$P_1 = \varphi_1' + \overline{\varphi_1'} + X, \quad D_1 = \bar{z}\varphi_1'' + \psi_1' + Y \tag{29}$$

Due to equilibrium the stress the functions $X$ and $Y$ are related to each other as

$$\partial_z X = \partial_{\bar{z}} Y \tag{30}$$

They can also be expressed through the function $W$ as follows

$$X = eP_0 + \left(\partial_z W + \partial_{\bar{z}} \overline{W}\right)\left(\kappa_0 - 1\right)^{-1}, \quad Y = eD_0 - \partial_z \overline{W} \tag{31}$$

Now by introducing new holomorphic functions

$$\varphi = \varphi_0 + \lambda\varphi_1, \quad \psi = \psi_0 + \lambda\psi_1 \tag{32}$$

one can represent the first approximation of the general solution in the form

$$2\mu_0 w = \kappa_0\varphi - z\overline{\varphi}' - \overline{\psi} + \lambda W, \quad P = \varphi' + \overline{\varphi}' + \lambda X, \quad D = \bar{z}\varphi'' + \psi' + \lambda Y \tag{33}$$

Formulae (33) can be considered as a generalization of the Kolosov formulae to the case of slightly non-homogeneous medium [10]. The obtained formulas are similar to the solution for homogeneous media with body forces. They would coincide if the first terms in both expressions (31) were omitted

# 3 Singular integral equations for crack problems

## 3.1 Reduction to the singular integral equation

Let us determine boundary values of the involved functions. Since on the boundary of a domain the complex variables are not independent, this makes it possible to consider the boundary value of the function as a function of a single variable

$$f(t) = f(t,\bar{t}) = \lim_{z \to t} f(z,\bar{z}), \quad t \in \Gamma \tag{34}$$

Here $\Gamma$ is the boundary of the domain, which can be considered as a union of open contours in the entire complex plane. For simplicity all the contours are assumed to be smooth, i.e. the do not contain any corner points.

The derivative of a complex valued function along the contour with respect to the complex coordinate $t$ has the form

$$f'(t) = \partial_z f\big|_{z=t,\bar{z}=\bar{t}} + e^{-2i\theta}\partial_{\bar{z}} f\big|_{z=t,\bar{z}=\bar{t}}, \quad t \in \Gamma \tag{35}$$

where is $\theta=\theta(t)$ the angle between the positive direction of tangent to the contour and the positive direction of the real axis.

Let us also introduce an operator $\Delta$, that being placed before any function $f$, denote a jump of this function across the contour

$$\Delta f(t) = f^+(t) - f^-(t) \tag{36}$$

where index "±" corresponds to the boundary values of the function on the upper/lower surface of $\Gamma$.

For the derivative of the complex displacements one finds

$$2\mu_0 w'(t) = (\kappa_0 + 1)\varphi'(t) - (P(t) + e^{-2i\theta}\overline{D}(t)) + \lambda W'(t), \tag{37}$$

Since the term in parentheses in the right hand side of (36) represent the complex vector of the applied load, $N+iT$, one can present the boundary value of the complex potential φ'(t) in the form

$$(\kappa_0 + 1)\varphi'(t) = 2\mu_0 w'(t) + N(t) + iT(t) - \lambda W'(t), \quad t \in \Gamma \tag{38}$$

In crack problems it is often accepted that the stress vector is continuous across the boundary, therefore $\Delta(N+iT)=0$. Bearing in mind this simplification and applying operator Δ to the both parts of (38) one finds the following expression for the jump of the sought complex potential

$$(\kappa_0 + 1)\Delta\varphi'(t) = 2\mu_0 \Delta w'(t) - \lambda\Delta W'(t), \quad t \in \Gamma \tag{39}$$

Then the potential can be found via the Cauchy integral

$$\varphi'(z) = \frac{1}{2i\pi}\int_\Gamma \frac{\rho(t)}{t - z}dt \tag{40}$$

where the unknown density $\rho(t)$ has the form

$$\rho(t) = \frac{2\mu_0 \Delta w'(t) - \lambda\Delta W'(t)}{\kappa_0 + 1} \tag{41}$$

It is evident that the expression (40) is the same as for homogeneous problem while the density contains an additional term, which is proportional to the jump of derivative of the particular solution $W$ that is known as soon as the solution of the problem for homogeneous case is known.

Once the function $\varphi'$ is known, the stress function $P$ can be expressed in the form

$$P = \frac{1}{2\pi i}\int_\Gamma \frac{\rho(t)}{t - z}dt - \frac{1}{2\pi i}\int_\Gamma e^{-2i\theta(t)}\frac{\overline{\rho(t)}}{\bar{t} - \bar{z}}dt + \lambda X \tag{42}$$

Stress function $D$ can be determined by the following way. It follows from the DEE that it can be sought in the form

$$D = \frac{-1}{2\pi i}\int_\Gamma \frac{(\bar{t} - \bar{z})\rho(t)}{(t - z)^2}dt + \omega + \lambda Y \tag{43}$$

where ω=ω(z) is an arbitrary holomorphic function. To find this function one needs to substitute (42) and (43) into the expression for the jump of the stress vector on Γ. Since it has been assumed that this jump is zero, by the Sokhotski-Plemelj formulae, [10], one should equate all non-integral terms to zero, which immediately fives the following expression for the jump of the holomorphic function, as follows

$$\Delta\omega(t) = -\overline{\rho(t)} - \lambda\Delta Q(t), \quad Q = X + Y \tag{44}$$

Then the holomorphic function ω(z) is determined by its jump, which eventually leads to the following expression for the stress function $D$

$$D = -\frac{1}{2\pi i}\int_{\Gamma}\frac{\overline{\rho(t)+\lambda\Delta Q(t)}}{t-z}dt - \frac{1}{2\pi i}\int_{\Gamma}\frac{(\bar{t}-\bar{z})\rho(t)}{(t-z)^2}dt + \lambda Y \quad (45)$$

As soon as the expressions for the stress functions via the unknown density $\rho(t)$ are found one can derive a singular integral equation (SIE) of the considered problem by assuming that the stress vector is known on Γ. The final expression for the SIE can be written in the form

$$\frac{1}{2\pi i}\int_{\Gamma}\left[\frac{\rho(t)}{t-\zeta} - e^{-2i\theta(t)}\frac{\overline{\rho(t)}}{\bar{t}-\bar{\zeta}}\right]dt + \\ +\frac{e^{-2i\theta(\zeta)}}{2\pi i}\int_{\Gamma}\left[\frac{\overline{\rho(t)+\lambda\Delta Q(t)}}{t-\zeta} + \frac{(\bar{t}-\bar{\zeta})\rho(t)}{(t-\zeta)^2}\right]dt = N + iT - \lambda Q, \quad \zeta \in \Gamma \quad (46)$$

The last term in the right hand side of (46) is known from the solution of the similar SIE for homogeneous domain in which it is necessary assume $W$=0 and $X$=$Y$=0. For a system of cracks SIE (46) can be transformed to a system of SIE, see [7] for detail.

It should be noted that in order to possess a unique solution the system (46) has to be complemented by the condition of single-valuedness of the displacements. This condition depends on the problem considered, for the case of $n$ non-intersecting cracks there should be specified $n$ additional conditions imposed in the sought function, which is similar to homogeneous case, see details in [11].

### 3.2 Particular case of a straight crack

In the case of a straight crack located on the interval (-1,1) of the real axis the SIE assumes simple form

$$\frac{1}{\pi i}\int_{-1}^{1}\frac{\overline{\rho(t)}}{t-x}dt = -[N(x)+iT(x)] + \lambda S(x), \quad |x|<1 \quad (47)$$

where the function $S(t)$ has been introduced as follows

$$S(x) = Q(x) - \frac{\Delta Q(x)}{2} - \frac{1}{2\pi i}\int_{-1}^{1}\frac{\Delta Q(t)}{t-x}dt \quad (48)$$

The condition of single valuedness assume the form

$$\int_{-1}^{1}[(\kappa_0+1)\rho(t) + \chi\Delta Q(t)]dt = 0 \quad (49)$$

 
www.witpress.com, ISSN 1743-355X (on-line)

The solution of this problem is obtained by perturbations, see [7] for detail. It has the form

$$\overline{\rho(t)} = \frac{i}{\pi\sqrt{1-t^2}}\left\{\int_{-1}^{1}\frac{\sqrt{1-x^2}}{x-t}[N(x)-iT(x)-\lambda S(x)]dx + C\right\} \tag{50}$$

An arbitrary constant $C$ should be determined from the condition of single-valuedness of the displacements which

It should be noted that functions $Q(t)$, $\Delta Q(t)$ and $S(x)$ in (47)-(50) remain unknown until the problem for the homogeneous medium is solved. This solution has the form

$$\overline{\rho_0(t)} = \frac{i}{\pi\sqrt{1-t^2}}\int_{-1}^{1}\frac{\sqrt{1-x^2}}{x-t}[N(x)-iT(x)]dx \tag{51}$$

Here the function $\rho(t)$ is decomposed into the sum

$$\rho(t) = \rho_0(t) + \lambda\rho_1(t) \tag{52}$$

where $\rho_0(t)=2\mu_0(1+\kappa_0)\Delta w'_0(t)$ represents a solution for the homogeneous plate

Now the function $Q(t)$ can be evaluated by substitution of the complex potential with zero sub-index in the corresponding formulae to determination of $W$, $X$, $Y$, $Q$ and $S$. Their explicit forms are not written here because of awkwardness.

By applying the perturbation method with the account for (52), the solution for $\rho_1(t)$ yields the form similar to (50) where the constant $C$ is to be determined from the condition (49) . This results in

$$C = \frac{-i\lambda}{\kappa_0+1}\int_{-1}^{1}\Delta Q(t)dt = 0 \tag{53}$$

Formulae (51)-(53) present the complete solution to the problem for the straight crack in slightly non-homogeneous plate.

# 4 Conclusions

The present paper presents a system of SIE for curvilinear non-intersecting cracks in slightly non-homogeneous media in the form (46). It can be solved analytically for some simple particular cases of straight or circular cracks. For curvilinear cracks or for a system of $n$ cracks the solution can be found by any numerical procedure associated with solving a system of SIE, in particular by the methods of mechanical quadratures [11].

The obtained system can be used for analysis of the characteristics related to the fracture of heterogeneous materials and in particular to investigation of the fracture propagation in FGM.

## References


[1] F. Delale, F. Erdogan, 1988. Interface crack in a non-homogeneous elastic medium. *Intt. J. Engng Sci.* **26**, 559-568.

[2] Y. F. Chen, F. Erdogan, 1996. The interface crack problem for a non-homogeneous coating bonded to a homogeneous substrate. *J. Mech. Phys. Solids*, **44**, 771-787.

[3] A.A. Al-Falou, R.C. Ball, 2000. Crack propagation in heterogeneous media. *Inte Journal of Solids and Structures* **37**, 461-475

[4] S. Roux, D. Vandembroucq, F. Hild, 2003. Effective toughness of heterogeneous brittle materials. *European Journal of Mechanics A/Solids* **22**, 743–749

[5] R. Luciano, J.R. Willis, 2003. Boundary-layer corrections for stress and strain fields in randomly heterogeneous materials. Journal of the Mechanics and Physics of Solids, **51**, 1075 – 1088.

[6] B.M. Singh, J. Rokne, J. Vrbik, R.S. Dhaliwal, 2006. Finite Griffith crack propagating in a non-homogeneous medium. *European Journal of Mechanics A/Solids,* **25**, 867–875

[7] Galybin, A.N., 1999. An analytic solution for a straight crack in slightly non-homogeneous elastic plate *Proc of Int Workshop on Fracture Mechanics and Advanced Engineering Materials* (Editors L. Ye and Y-W Mai). The University of Sydney, Sydney, Australia. pp. 68-75.

[8] Vekua, I.N. 1962. *Generalised analytic functions.* Pergamon Press, Addison Wesley Publishing Company inc. Reading, Massachusetts, UWA.

[9] Vekua, I.N. 1967. *New methods for solving elliptic equations.* North-Holland Publishing Company. John Wiley and Sons Inc. New York, UWA.

[10] Muskhelishvili, N.I. (1953) *Some Basic Problems of the Mathematical Theory of Elasticity.* P. Noordhoff Ltd. Groningen-Holland.

[11] Savruk, M.P. 1981. Two-Dimensional Problems of Elasticity for Bodies with Cracks. Kiev, Naukova Dumka (in Russian).

# A simple implementation of the dual boundary element method using the tangential differential operator for plane problems

L. Palermo, P. C. Gonçalves & L. G. Figueiredo
*Faculty of Civil Engineering, Architecture and Urban Design, State University of Campinas, Brazil*

## Abstract

Plane problems of linear fracture mechanics were solved with the dual boundary element method (DBEM) using the tangential differential operator (TDO) in traction boundary integral equation (BIE). The numerical implementation employed same shape functions for conformal and non-conformal interpolations with nodal parameters fixed at the ends of elements. Different collocation point positions were used in crack surfaces according to continuity requirements related to each BIE type employed. The aim of this paper is presenting a simple implementation for fracture analysis. Results obtained in the literature for stress intensity factors were included to show the accuracy of the formulation.
*Keywords: tangential differential operator, dual boundary element method, stress intensity factor.*

## 1 Introduction

This study began when the implementation presented in [1] was revisited in [2] with purpose to apply conformal interpolations along crack surfaces. Non-conformal interpolations along crack surfaces were used in [1] with quadratic elements and collocation points positioned at element nodes. The problems solved in [1] could be analyzed in [2] using conformal interpolations along crack surfaces with collocations points shifted to the interior of elements. The obtained results in [2] with quadratic or linear elements had a good agreement with those presented in [1] for internal positions ($\xi'$) of collocation points equal $\pm 0.67$ in the range (-1, 1). Furthermore, results were not significantly changed if conformal or

WIT Transactions on Modelling and Simulation, Vol 50, © 2010 WIT Press
www.witpress.com, ISSN 1743-355X (on-line)
doi:10.2495/BE100071

non-conformal interpolation were applied along crack surfaces because their shape functions employed same expressions and nodal parameters were always fixed at the ends of elements. Traction BIE using the tangential differential operator (TDO) was studied in [3] with shape functions and collocation point positions studied in [2]. The order of the strong singularity was reduced with the operator and the displacement function derivative was adopted in the kernel containing TDO. Furthermore, an additional term related to TDO was introduced to allow discontinuities in the boundary line usually due to non-conformal interpolations at crack tips and corners when conformal interpolations were applied along crack surfaces. The results were not degenerated in [3] with reference to [1] or [2] even using conformal interpolations along crack surfaces and isoparametric linear elements.

Internal positions for collocation points were used in [1–3] for both BIEs of DBEM and the present study analyzes if collocation points of both BIEs should have same positions in opposite elements at crack surfaces when conformal interpolations are adopted or different positions could be applied.

## 2 The dual boundary element method using the tangential differential operator

The BIE for the gradient at an internal point x is written as follows using the differentiation in terms of field variables:

$$u_{i,m}(x) = \int_{\Gamma} T_{ij,m}(x,y)\,u_j(y)\,d\Gamma(y) - \int_{\Gamma} U_{ij,m}(x,y)\,t_j(y)\,d\Gamma(y) \qquad (1)$$

$U_{ij}(x, y)$ and $T_{ij}(x, y)$ are the displacement and the traction, respectively, in direction j at the boundary point y, due to a singular load in direction i at collocation point x', according to the Kelvin solution for two-dimensional problems; $u_j(y)$ and $t_j(y)$ are the displacement and the traction at the field point, respectively.

After some algebraic manipulation and employing properties of Kelvin type fundamental solutions, the TDO is obtained from integration by parts [3, 4] and the BIE for the gradient at an internal point x with TDO is given by:

$$u_{i,m}(x) = \int_{\Gamma} \sigma_{ibj}(x,y)\,D_{mb}\left[u_j(y)\right]d\Gamma(y) - \int_{\Gamma} U_{ij,m}(x,y)\,t_j(y)\,d\Gamma(y) \qquad (2)$$

$D_{bm}(\,)$ is the tangential differential operator, which has the following definition:

$$D_{mb}\left[u_j(y)\right] = n_m(y)\,u_{j,b}(y) - n_b(y)\,u_{j,m}(y)$$

The integrals of eq. (2) are regular for internal points and exhibit singularities of order 1/r when the field point approaches the collocation point. Stress BIE is obtained from eq. (1) or eq. (2) using the Hooke tensor and the symmetry property of $U_{ij,m}(x, y)$. The limiting form of stress BIE at an internal point when it is led to a point on the boundary defines stress BIE at a boundary point, which has the following expression at point x' on a smooth boundary:

$$\frac{1}{2}\sigma_{ak}(x') = C_{aki\,m}\int_{\Gamma} T_{ij,m}(x',y)u_j(y)d\Gamma(y) - \int_{\Gamma}\sigma_{jak}(x',y)t_j(y)d\Gamma(y) \quad (3)$$

$$\frac{1}{2}\sigma_{ak}(x') = C_{aki\,m}\int_{\Gamma}\sigma_{ibj}(x',y)D_{mb}\left[u_j(y)\right]d\Gamma(y) - \int_{\Gamma}\sigma_{jak}(x',y)t_j(y)d\Gamma(y) \quad (4)$$

$$C_{akim} = \mu\left(\frac{2\nu}{1-2\nu}\delta_{ak}\delta_{im} + \delta_{ai}\delta_{km} + \delta_{am}\delta_{ki}\right)$$

$C_{akim}$ is the Hooke tensor for isotropic media, $\nu$ is the Poisson ratio and $\mu$ is equal to the shear modulus. $\delta_{ij}$ is the Kronecker delta

An additional term is necessary in equation (4) when a discontinuity in the boundary line is introduced due to a displacement discontinuity:

$$\begin{aligned}\frac{1}{2}\sigma_{ak}(x') = {} & C_{aki\,m}\int_{\Gamma}\sigma_{ibj}(x',y)D_{mb}\left[u_j(y)\right]d\Gamma(y) + \\ & + \left[e_{3bm}\sigma_{ibj}(x',y)u_j(y)\right]_0^{\Gamma} - \int_{\Gamma}\sigma_{jak}(x',y)t_j(y)d\Gamma(y)\end{aligned} \quad (5)$$

$e_{3bm}$ is the permutation symbol.

Traction BIE is obtained from eq. (3) or (5) when the stress tensor at boundary point x' is multiplied by direction cosines of the outward normal at this point ($n'_a$). The corresponding traction BIEs from eq. (3) and (5) are given by:

$$\frac{1}{2}t_k(x') = n'_a(x')C_{aki\,m}\int_{\Gamma} T_{ij,m}\,u_j\,d\Gamma - n'_a(x')\int_{\Gamma}\sigma_{jak}\,t_j\,d\Gamma \quad (6)$$

$$\begin{aligned}\frac{1}{2}t_k(x') = {} & n'_a(x')C_{aki\,m}\int_{\Gamma}\sigma_{ibj}\,D_{mb}(u_j)d\Gamma + \\ & + n'_a(x')\left[e_{3bm}\sigma_{ibj}u_j\right]_0^{\Gamma} - n'_a(x')\int_{\Gamma}\sigma_{jak}\,t_j\,d\Gamma\end{aligned} \quad (7)$$

General mixed-mode crack problems can be solved with a single domain formulation in DBEM when the traction BIE (eq. (6) or (7)) is applied to one of the crack surfaces and the displacement BIE to the other. Although the integration path is still the same for coincident points on the crack surfaces, the respective boundary integral equations are now distinct. The displacement BIE is given by:

$$\frac{1}{2}u_i(x') + \int_{\Gamma} T_{ij}(x',y)u_j(y)d\Gamma(y) = \int_{\Gamma} U_{ij}(x',y)t_j(y)d\Gamma(y) \quad (8)$$

The collocation points must be positioned to satisfy the continuity requirements for each BIE. Continuity of the displacement function at x' is the necessary condition for displacement BIE and it is satisfied when the collocation point is placed at the ends of the boundary element or at its interior. Continuity of the displacement derivative at x' is required for traction BIE and it is satisfied when the collocation point is placed at the interior of the element [1, 4].

# 3 Numerical implementation

The integrals of eq. (6) exhibit singularities of order 1/r and $1/r^2$. The strong singularity ($1/r^2$) requires the use of the Cauchy and the Hadamard principal-value integral. The application of second order finite part [1] results in analytical expressions for singular terms and a numerical integration using the Gauss-Legendre scheme for the regular term. Hadamard principal value integral is not used when traction BIE employs TDO because integrals of eq. (7) exhibit singularity types of order 1/r like in eq. (8). The improper integral only requires the use of the Cauchy principal value sense. Analytical expressions can be used to evaluate singular integrals with the Cauchy principal value sense and Gauss-Legendre scheme used for regular integrals. The present numerical implementation for terms containing the tangential differential operator considered the differentiation of the displacement function [5] and the following relation was applied:

$$D_{bm}[f(y)] = n_b(y) f_{,m}(y) - n_m(y) f_{,b}(y) = e_{3\gamma\rho} \frac{1}{|J|} \frac{df}{d\xi}$$

J is Jacobian of the transformation and $\xi$ is the intrinsic coordinate to perform the integration on the element; f corresponds to the function representing displacement in elements.

Linear shape functions were used to represent displacements and tractions in boundary elements. Same shape functions were used for conformal and non-conformal interpolations with nodal parameters positioned at ends of elements.

Collocation points for displacement BIE were positioned at nodes of elements in conformal interpolation and shifted to the interior of the element in non-conformal interpolation with positions ($\xi'$) equal $\pm 0.67$ in the range (-1, 1). On the other hand, collocation points for traction BIE were always shifted to the interior of elements and the positions ($\xi'$), in range (-1, 1), were: i) $\xi' = -0.67$ for continuous elements; ii) $\xi' = -0.67$ and $\xi' = +0.67$ for discontinuous elements.

The diagonal terms were directly obtained using the shape function and the collocation point position in the element.

# 4 Stress intensity factor evaluation

Stress intensity factors were assessed using the near-tip displacement extrapolation [1]. Considering a polar coordinate system (r, $\theta$) centered at the crack tip, such that crack surfaces could be defined with $\theta = \pm \pi$. The displacement field on the crack surface has the following expressions considering the first term of William's expansion:

$$u_2(\theta = \pi) - u_2(\theta = -\pi) = \frac{\kappa + 1}{\mu} K_I \sqrt{\frac{r}{2\pi}} \tag{9}$$

$$u_1(\theta = \pi) - u_1(\theta = -\pi) = \frac{\kappa + 1}{\mu} K_{II} \sqrt{\frac{r}{2\pi}} \tag{10}$$

The stress intensity factors for deformation modes I and II are $K_I$ and $K_{II}$, respectively; the parameter κ is equal to 3-4η; η is equal to ν for plane strain problems and equal to ν/(1+ν) for plane stress problems The near-tip displacement extrapolation works with eq. (9) and (10) to obtain stress intensity factors when the displacements are known. The situation is shown in Figure 1, where opposite elements share the crack tip at nodes B and C. The length of the linear element is equal to l. The stress intensity factors are given by:

$$K_I^{DE} = \left(u_2^D - u_2^E\right)\frac{\mu}{\kappa+1}.\sqrt{2}.\sqrt{\frac{\pi}{l}} \quad (11)$$

$$K_{II}^{DE} = \left(u_1^D - u_1^E\right)\frac{\mu}{\kappa+1}.\sqrt{2}.\sqrt{\frac{\pi}{l}} \quad (12)$$

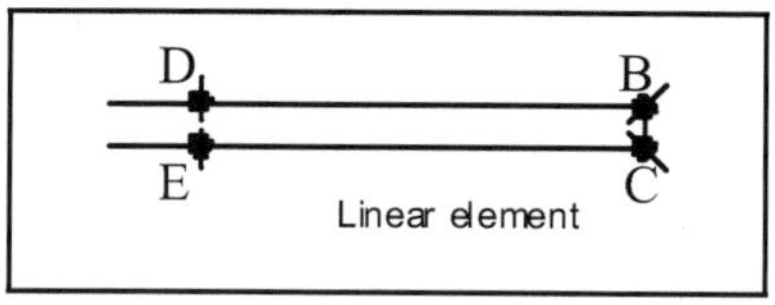


Figure 1: Crack tip at points B and C.

## 5 Numerical examples

A rectangular plate containing a single horizontal edge crack shown in Figure 2 used a mesh with 48 linear elements plus 8 elements on each crack surface (64 B.E.). The crack length is a, the plate width is w and the height is 2h. A uniform traction in the height direction was symmetrically applied at the ends. Results obtained for the ratio h/w equals to 0.5 are shown in Table 1. Three ratios a/w were considered: 0.2, 0.4, and 0.6. Stress intensity factor were obtained with eq. (11).

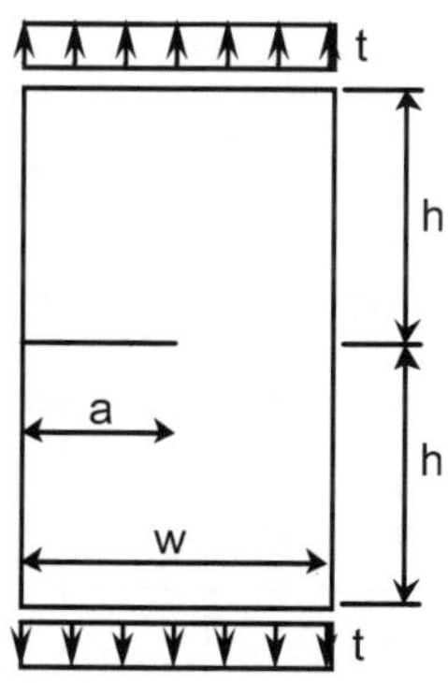


Figure 2: Single edge crack.

Table 1: Single edge crack with 64 linear elements: $K_I/(t\sqrt{\pi a})$.

| a/w | [6] | [1] | [3] | Obtained |
|---|---|---|---|---|
| 0.2 | 1.488 | 1.566 | 1.501 | 1.642 |
| 0.4 | 2.324 | 2.230 | 2.368 | 2.328 |
| 0.6 | 4.152 | 4.580 | 4.254 | 4.209 |

A rectangular plate containing a central slant crack shown in Figure 3 had a mesh with 48 linear elements plus 12 elements on each crack surface (72 B.E.). Crack length is 2a, the plate width is 2w and the height is 2h. A uniform traction in the height direction was symmetrically applied at the ends. Results obtained for h/w ratio equal to 2 are shown in Table 2 for $K_I$ and in Table 3 for $K_{II}$. Three a/w ratios were considered: 0.2, 0.4, and 0.6. Stress intensity factors were obtained with eq. (11) and eq. (12).

A rectangular plate containing an internal kinked crack shown in Figure 4 had a mesh with 60 linear elements plus 10 elements on each horizontal crack surface and 8 elements on each inclined crack surface. One of the segments of the crack is horizontal with length b while the other segment makes an angle of 45 degrees with the horizontal and has length a; the horizontal projection of the total crack is given by $2c=b+a(\sqrt{2})/2$. The kink of the crack is at the center of the plate, the plate width is 2w and the height is 2h. Three a/b ratios were considered: 0.2, 0.4 and 0.6. The results obtained for b/w equal to 0.1 are shown in Tables 4 to 7. Stress intensity factors were obtained with eq. (11) and eq. (12).

Table 2: Central slant crack with 72 linear elements: $K_I/(t\sqrt{\pi a})$.

| a/w | [7] | [1] | [3] | Obtained |
|---|---|---|---|---|
| 0.2 | 0.518 | 0.531 | 0.513 | 0.512 |
| 0.4 | 0.572 | 0.588 | 0.567 | 0.566 |
| 0.6 | 0.661 | 0.686 | 0.660 | 0.658 |

Table 3: Central slant crack with 72 linear elements: $K_{II}/(t\sqrt{\pi a})$.

| a/w | [7] | [1] | [3] | Obtained |
|---|---|---|---|---|
| 0.2 | 0.507 | 0.519 | 0.502 | 0.501 |
| 0.4 | 0.529 | 0.541 | 0.523 | 0.522 |
| 0.6 | 0.567 | 0.579 | 0.561 | 0.557 |

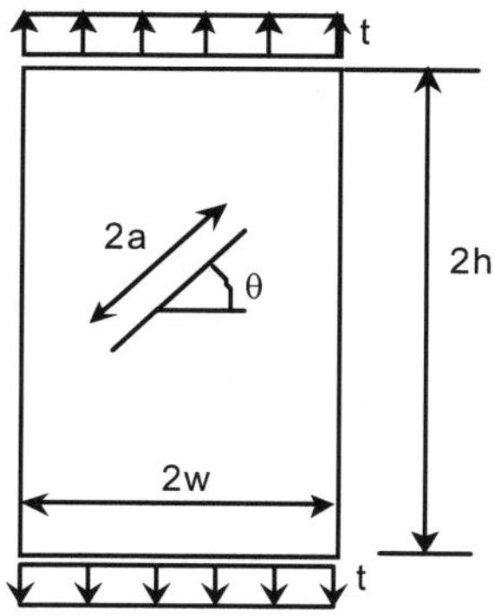


Figure 3: Central slant crack.

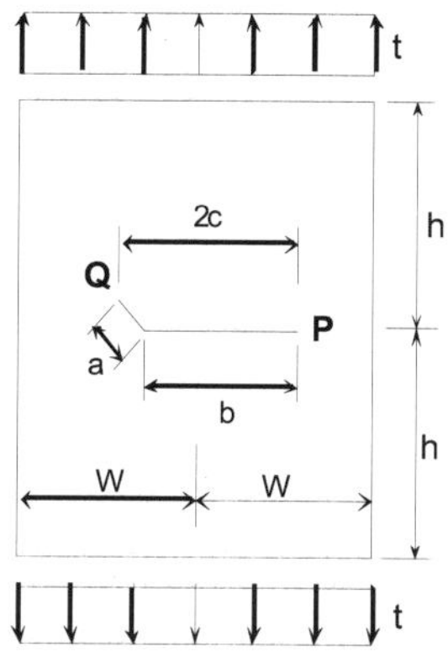


Figure 4: Central kinked crack.

Table 4: Kinked crack with 96 linear elements, results at P: $K_I/(t\sqrt{\pi a})$.

| a/w | [7] | [1] | [3] | Obtained |
|---|---|---|---|---|
| 0.2 | 0.995 | 1.021 | 0.988 | 0.988 |
| 0.4 | 0.990 | 1.018 | 0.985 | 0.984 |
| 0.6 | 0.986 | 1.017 | 0.983 | 0.982 |

The obtained values were close to the literature. The adopted meshes were similar to those suggested in [1]. Conformal interpolations were used along the straight lines shown in Figures 2 to 4. Double nodes were introduced at corners and at crack tips. Collocation points for displacement BIE were positioned at element nodes in conformal interpolations and shifted to the interior of element for double nodes at corners and crack tips. Collocation points for traction BIE were always positioned inside the elements. Internal positions for collocations point were studied in [8] for the present strategy and the best position (±0.67), which was adopted in this study, was equal to that obtained in [2] and used in [3]. By the way, when a/w is equal to 0.2, values for $K_I$ in Table 1 and in Table 6 as well as for $K_{II}$ in Table 7 could be improved if ξ' could be equal to ±0.75.

Table 5: Kinked crack with 96 linear elements, results at P: $K_{II}/(t\sqrt{\pi a})$.

| a/w | [7] | [1] | [3] | Obtained |
|---|---|---|---|---|
| 0.2 | 0.028 | 0.030 | 0.029 | 0.028 |
| 0.4 | 0.033 | 0.036 | 0.035 | 0.035 |
| 0.6 | 0.030 | 0.032 | 0.032 | 0.032 |

Table 6: Kinked crack with 96 linear elements, results at Q: $K_I/(t\sqrt{\pi a})$.

| a/w | [7] | [1] | [3] | Obtained |
|---|---|---|---|---|
| 0.2 | 0.598 | 0.634 | 0.636 | 0.630 |
| 0.4 | 0.574 | 0.603 | 0.606 | 0.603 |
| 0.6 | 0.568 | 0.595 | 0.600 | 0.595 |

Table 7: Kinked crack with 96 linear elements, results at Q: $K_{II}/(t\sqrt{\pi a})$.

| a/w | [7] | [1] | [3] | Obtained |
|---|---|---|---|---|
| 0.2 | 0.557 | 0.589 | 0.590 | 0.591 |
| 0.4 | 0.607 | 0.637 | 0.639 | 0.635 |
| 0.6 | 0.627 | 0.659 | 0.661 | 0.655 |

## 6 Conclusions

The present implementation for DBEM considered conformal interpolation along crack surfaces using same expressions for shape functions of conformal and non-conformal interpolations with nodal parameter always fixed at the ends of elements. Internal positions for collocations points were used to non-conformal interpolations for displacement BIE or to all interpolation types for traction BIE. This implementation can be understood as a current code for displacement BIE improved with the traction BIE formulation to treat linear fracture mechanics problems, i.e. no special care was necessary in the implementation to introduce the displacement BIE to treat linear fracture mechanics.

Thus, the effect of the collocation point position is only related to the collocation point for traction BIE. The present results were similar to those obtained with internal points for both BIEs in [3] as well as to those obtained with quadratic elements using non-conformal interpolations along crack surfaces in [1].

The stress intensity factors were obtained using near-tip displacement extrapolation and it is important to note that better values for them could be obtained with J-integral technique [1], which is a post processing technique that could be applied to the present formulation but was not used in this paper.

This study strongly suggests the use of the TDO due to the simplifications to treat the singularity (the treatment of the improper integrals only used the first order finite part). Furthermore, the use of derivatives of the adopted shape function for displacement without using other interpolation for TDO was an interesting alternative. It is important to note that constant values were obtained as derivatives of the linear shape function and the results were not degraded.

## Acknowledgement

The author is grateful to CNPq for support for the development of the present research.

## References

[1] Portela, A, Aliabadi, MH, Rooke, DP, The dual boundary element method: Effective implementation for crack problems, International Journal of Numerical Methods in Engineering, 33, 1269-1287, 1992.

[2] Almeida, L.P.C.P.F., Palermo Jr., L, On the Implementation of the Two Dimensional Dual Boundary Element Method for Crack Problems, "5th International Conference on Boundary Elements Techniques" Lisboa, Portugal, 2004.

[3] Palermo Jr., L., Almeida, L.P.C.P.F. and Gonçalves, P.C., The Use of the Tangential Differential Operator in the Dual Boundary Element Equation, Structural Durability & Health Monitoring, vol.2, no.2, pp.123-130, Tech Science Press, 2006.

[4] Bonnet, M. Boundary Integral Equation Methods for Solids and Fluids, John Wiley & Sons Ltd, 1999.

[5] Kane, J. H., Boundary Element Analysis in Engineering Mechanics. Prentice Hall, 1994.

[6] Civelek, M. B., Erdogan, F., 1982, Crack problems for a rectangular sheet and an infinite strip, International Journal of Fracture, vol.19, pp.139-159.

[7] Murakami, Y., 1987, Stress Intensity Factors Handbook, Pergamon Press, Oxford.

[8] Palermo Jr., L., Gonçalves, P. C., Figueiredo, L. G., The collocation point position in the dual boundary element equation with the tangential differential operator, 19th International Congress of Mechanical Engineering, Brasília, Brasil, 2007.

# BEMs to evaluate interface cracks

P. Brož
*Professional Association for Science, Research and Consultancy, Czech Republic*

## Abstract

Boundary element approach to the 3D analysis of bimaterial interface cracks is put on. *J*-integral and stress intensity factors are calculated along the crack face employing the Energy Domain Integral and the $M_I$-integral techniques. The set-up numerical means is employed to analyze the problem of a fibre/matrix interface crack subject to lateral loading so as to evaluate the factors of the issues obtained from two-dimensional simulations. The obtained outcomes demonstrate the important role played by the elastic properties of the fibre, the matrix and the laminate in the mixed mode fracture condition. Numerical tests performed indicate that the multi-domain procedure is suitable for bimaterial stress analysis, and as well as competent to yield precise results of the stress intensity factors $K_{\mathrm{I}}$ and $K_{\mathrm{II}}$. The multi-domain boundary contour method with possible applications to interface and dissimilar material problems and also a technique tackling with a general type of displacement and traction compatibility conditions are introduced.

*Keywords: bimaterial plate, debonding, energy domain integral, interface crack, normalized SIF, shared node, virtual crack extension.*

## 1 Introduction

Composite materials have highly anisotropic properties, with superb stiffness and strength behaviour in the fibre direction and rather weak properties in the transverse direction. For this reason, laminates are used being made up of several stacked plies with different fibre orientations, causing sufficient stiffness in more than one direction. Nevertheless, mechanical loading of such structure also induces loads in arbitrary directions not the same either with those of the fibre or the applied load. The reason of early failure of the transverse ply is the material heterogeneity. It has been demonstrated that a global strain of 1% already

WIT Transactions on Modelling and Simulation, Vol 50, 
www.witpress.com, ISSN 1743-355X (on-line)
doi:10.2495/BE100081

produces local strains of more than 5%. An irregular packing of fibres owing to the inhomogeneous fibre distribution, increases the effect. Due to this local stress concentration, the macroscopic transverse failure strain is, in the majority of cases, extensively lower than the longitudinal failure strain. In all loading directions, the macroscopic characteristics of unidirectional fibre-reinforced-composite materials are hard affected by the events being present at the fibres scale. In longitudinal as well as in transverse direction, the stress transfer from the matrix to the fibres determines the mechanical behaviour of the composite, for the failure process is principally established by the strength of the fibre/matrix adherence. The stress concentrations around the fibres end in micro-crack initiation that will grow along a ply to come to be a complete lateral crack.

Recently, broad analyses have been performed and many questions regarding the mechanics of interface fracture have been solved. But still, progress has been generally oriented on the two-dimensional idealization of an interface crack, and restricted endeavour has been conducted on the three-dimensional concept of interface fracture. This is partly owing to the extreme complexity of such problems and very large computational efforts required for their numerical analysis. However, given the material mismatch at the interface boundary, it goes without saying that the three-dimensional effect plays a more outstanding role in a biomaterial structure than in a homogenous one.

Due to the inherent characteristics of the BEM formulation, it provides very precision issues for problems including strong geometrical discontinuities. This makes BEM a powerful numerical means to simulate crack problems.

BEM has been far-reachingly employed to analyze an amount of problems involving two-dimensional interface cracks [1, 2]. There are yet no reported results for three-dimensional problems, with the exception of an in-depth and superior study [3]. Though many authors suggest displacement and stress extrapolation methods to stipulate stress intensity factors from BEM outcomes [4, 5], $J$-integral methods make up a more strong approach. As an energy concept, $J$-integral methods eliminate the need to solve local crack tip fields precisely, because if integration domains are defined over a relatively large part of the mesh, accurate modelling of the crack tip is superfluous, since the contribution to $J$-integral of the crack tip fields is not meaningful. At the same time, BEM is especially appropriate for the evaluation of path independent integrals, as the required stresses, strains and derivatives of displacements at internal points can be directly obtained from their boundary integral representations. It has been also shown that BEM yields more accurate stresses and strains at internal points when comparing with other numerical techniques, therefore better results can be achieved. Use of the $J$-integral methodology for two-dimensional interface cracks can be searched in the work by Miyazaki et al. [6].

From the available algorithms for the numerical computation of the $J$-integral in three dimensions, the Energy Domain Integral (EDI) has been frequently used to employ. Together with the EDI, the interaction or $M_I$-integral methodology due to Chen and Shield [7] is used in this work, in compliance with [3], for decoupling the $J$-integral into the mixed–mode stress intensity factors. The $M_I$-

integral methodology is grounded on the superposition of two equilibrium states, formulated by the current problem and a set of auxiliary known solutions. The procedure has been recently announced in a number of papers using FEM to calculate stress intensity factors along three-dimensional interface cracks [8, 9]. In the field of fracture mechanics the solution to the applications of very interest is facilitated by using a multi-domain technique grounded on the boundary contour method [15].

## 2 *J*-integral and stress intensity factor calculation

Take a three-dimensional crack face with a continuously turning tangent as depicted in Fig. 2(a). Characterize a local coordinate system $x^*$ at position $\eta$, where the crack energy release rate is evaluated, given by $x_1^*$ normal to the crack front, $x_2^*$ normal to the crack plane, and $x_3^*$ tangent to the crack front.

The energy release rate $G(\eta)$ owing to crack extension in its own plane along a three-dimensional crack front reads

$$G(\eta)=\lim_{C\to 0}\xi_k(\eta)\int_{C(\eta)}\left(w\cdot\delta_{ki}-\sigma_{ij}^* u_{j,k}^*\right)n_i\,\mathrm{d}C \tag{1}$$

where $w$ is the strain energy density, $\sigma_{ij}^*$ and $u_{j,k}^*$ are Cartesian components of stress and displacement derivatives expressed in the local system $x^*$, $\xi_k(\eta)$ is the unit outward normal to the crack front in the local crack plane $x_1^*-x_3^*$ , $n_i$ is the unit vector normal to the contour $C(\eta)$ (which lies in the $x_1^*-x_2^*$ plane), and $\mathrm{d}C$ is the differential of the arc length $C$. Although eqn (1) comes from a two-dimensional analysis, it applies for the three-dimensional case, in the limit as $C\to 0$, plain strain conditions predominate three-dimensional fields approach to the plane problem.

So as to deduce the equivalent domain representation of eqn (1), we consider a small segment $L_c$ of the crack front that lies in the local $x_1^*-x_3^*$ plane. Next we suppose that the segment undergoes a virtual crack advance in the plane of the crack, and we define the magnitude of the advance at each point $\eta$ as $\Delta a(\eta)$. We note that $\Delta a(\eta)$ varies continuously along $L_c$ and disappears at each end of the segment. We have

$$G(\eta)=\int_{L_c} G(\eta)\Delta a(\eta)\mathrm{d}\eta \tag{2}$$

where $G(\eta)$ is the integral defined in eqn (1). When $G(\eta)$ is part of the point-wise energy release rate, $\overline{G}$ gives the total energy released when the finite segment $L_c$ experinces the virtual crack advance.

The relevant domain shape of the point-wise crack-tip contour integral can be obtained from eqn (2) by considering a tubular domain $V$ surrounding the crack segment (see Fig. 3). As shown in the figure, the surface $S_t$ is formed by translating the contour $C$ along the segment $L_c$, and $S_o$ stands for the outer surface of $V$ including the ends. Next an auxiliary function $q$ is introduced, which is sufficiently smooth in $V$ and it is determined on the surfaces of $V$ in the form:

$$q_k = \begin{cases} \Delta a(\eta) \cdot \xi_k(\eta), & \text{on } S_t, \\ 0, & \text{on } S_o \end{cases} \tag{3}$$

Lastly, in the limit as the tubular surface $S_t$ is thrusted onto the crack segment $L_c$ and in the absence of crack face tractions, we obtain the domain integral:

$$\overline{G} = \int_V \left(\sigma_{ij}^* u_{j,k}^* - w \cdot \delta_{ki}\right) q_{k,i} \, \mathrm{d}V \tag{4}$$

During absence of body forces the integral $\overline{G}$ according to eqn (4) reduces to the domain representation of the familiar $J$-integral. If it is supposed that $G(\eta)$ is constinuous along $L_c$, at once it follows from eqn (2) that:

$$J(\eta) = G(\eta) = \frac{\overline{G}}{\int_{L_c} \Delta a(\eta) \mathrm{d}\eta} \tag{5}$$

We put on the interaction or $M_1$-integral methodology for decoupling three-dimensional mixed-mode stress intensity factors in bimaterial interface cracks. So as to make sure the purpose of the stress intensity factors we must consider the linear elastic solution of the *open model* of interface cracks. It is assumed in this model that the small contact zone that always develops at the crack tip is physically non-relevant. We label the material occupying the upper half-plane as material 1 with Young modulus $E_1$ and Poisson ratio $\nu_1$ (see Fig. 1). The material occupying the lower half-plane has Young modulus $E_2$ and Poisson ratio $\nu_2$. Let us consider now two equilibrium states with field variables denoted by the superscripts (1) and (2). Superposition of the two equilibrium states results in another one, (1 + 2). Next, the stress intensity factors $K_j^{(1+2)}$ are equal to

$$K_j^{(1+2)} = K_j^{(1)} + K_j^{(2)} \quad (j = \mathrm{I, II, III}) \tag{6}$$

The relationship between the $J$-integral and the stress intensity factors of an interface crack may be written in the form

$$J = \frac{1}{E^* \cosh^2(\pi\varepsilon)} \left[K_I^2 + K_{II}^2\right] + \frac{1}{2\mu^*} K_{III}^2 \tag{7}$$

where $E^*$ and $\mu^*$ are the effective Young and shear modulus

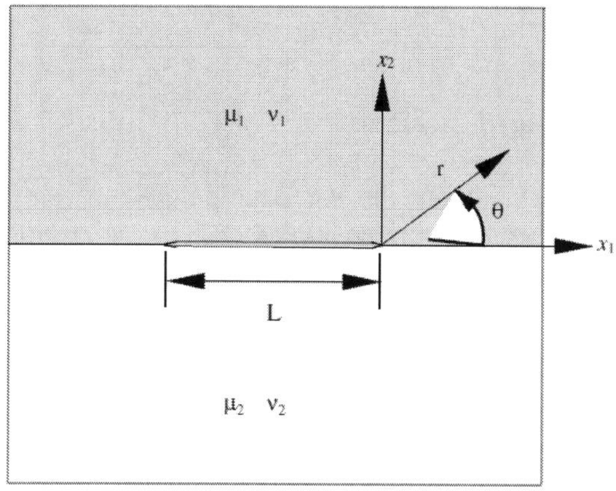


Figure 1: Bimaterial plate weakened by an interface crack (taken from [3]).

$$\frac{1}{E^*}=\frac{1}{2}\left(\frac{1-\nu_1^2}{E_1}+\frac{1-\nu_2^2}{E_2}\right),\ \frac{1}{\mu^*}=\frac{1}{2}\left(\frac{1}{\mu_1}+\frac{1}{\mu_2}\right) \tag{8}$$

and $\varepsilon$ represents the bimaterial constant

$$\varepsilon=\frac{1}{2\pi}\left(\frac{\kappa_1\mu_2+\mu_1}{\kappa_2\mu_1+\mu_{21}}\right) \tag{9}$$

According to eqns (6), (7), the stress intensity factors can be related to the $J$-integral for the superimposed state (1 + 2) leading to

$$J^{(1+2)}=\frac{1}{E^*\cosh^2(\pi\varepsilon)}\left[\left(K_I^{(1+2)}\right)^2+\left(K_{II}^{(1+2)}\right)^2\right]+\frac{1}{2\mu^*}\left(K_{III}^{(1+2)}\right)^2 \tag{10}$$

We may rewrite eqn (10) by virtue of the stress intensity factors for the equilibrium states (1) and (2), to pass:

$$J^{(1+2)}=J^{(1)}+J^{(2)}+\frac{2}{E^*\cosh^2(\pi\varepsilon)}\left[K_I^{(1)}K_I^{(2)}+K_{II}^{(1)}K_I^{(2)}\right]+\frac{1}{2\mu^*}K_{III}^{(1)}K_{III}^{(2)} \tag{11}$$

Then, the $M_I$-integral is determined being

$$M_1=J^{(1+2)}-J^{(1)}-J^{(2)}=\frac{2}{E^*\cosh^2(\pi\varepsilon)}\left[K_I^{(1)}K_I^{(2)}+K_{II}^{(1)}K_I^{(2)}\right]+\frac{1}{2\mu^*}K_{III}^{(1)}K_{III}^{(2)} \tag{12}$$

In compliance with eqn (4), a domain representation of the $M_I$-integral can be obtained in the following form:

$$M_1 = \int_V \left(\sigma_{ij}^{*(1)} u_{j,k}^{*(2)} + \sigma_{ij}^{*(2)} u_{j,k}^{*(1)} - \sigma_{ij}^{*(1)} \varepsilon_{ij}^{*(2)} \delta_{ki}\right) q_{k,i} \, \mathrm{d}V \tag{13}$$

# 3 Boundary element realization

Relations (4) and (13) facilitate the computation of $J$-integral and the mixed-mode stress intensity factors at any position $\eta$ on the crack front to be performed. In each instance, this requires the evaluation of a volume integral within closed domains that enclose a segment of the crack front $L_c$. A natural choice here is to make $\eta$ coincident with the element nodes on the crack front, while $L_c$ is taken as the element or element sides at which points $\eta$ lie (see Fig. 2). The portion of the model domain in which the volume integrals are evaluated is discretized using 27-noded isoparametric (brick) cells, over which stresses, strains and displacements derivatives are approximated by products of the cell interpolation functions $\Psi_i$ and the nodal values of $\sigma_{ij}$, $\varepsilon_{ij}$ and $u_{i,j}$. Nodal values of this variable depend on whether the node is internal or lies on the model boundary. Volume discretization is designed to have web-style geometry around the crack front, while the integration volumes are taken to coincide with the different rings of cells, what is demonstrated in Fig. 3, where the front of the model was partially removed to point out the crack and the integration domains.

Conforming to Fig. 2, three different cases are in want of the reflection, depending on whether the node of interest M is in the middle of an element side (mid-side node), it is shared by two elements (shared node), or it is situated the same with the external surface (surface node). If the node $M$ is a mid-side node or surface node, $L_c$ (the segment of the crack front over which the $M_1$-integral is computed) spans over one element, connecting nodes $M-1$, $M$, and $M+1$ and nodes $M-2$, $M-1$ and $M$, respectively. On the other hand, if $M$ is a shared node, $L_c$ spans over two elements, connecting nodes from $M-2$ to $M+2$.

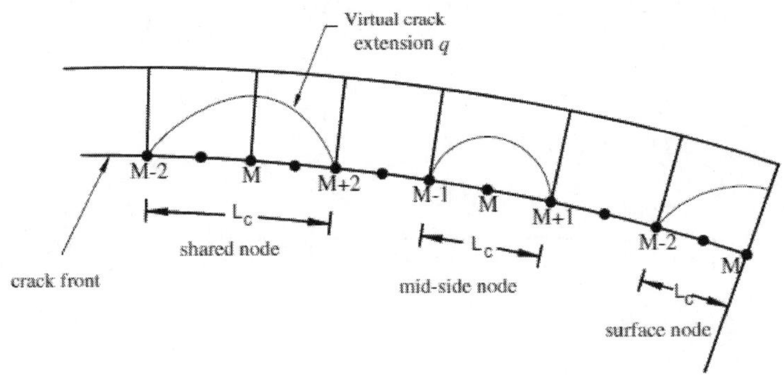


Figure 2: Scheme of the volume cells in the crack front region explaining the virtual crack stretches for a shared node, a mid-side node and a surface node [3].

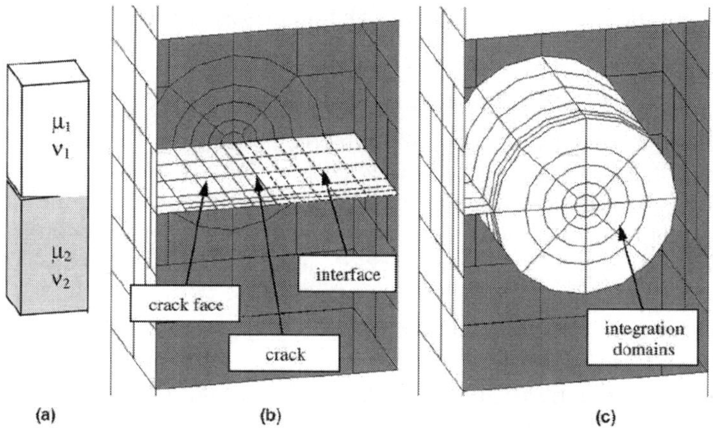


Figure 3: (a) Problem geometry, (b) boundary element discretization (c) integration domains.

# 4 Explanation of a fibre/matrix interface crack subject to lateral loading

A fibre/matrix interface crack is a special example of interface crack. High-resolution numerical BEM models [11] and analytical solutions [12] have shown that for arc-shaped debonds an extensive region with negative opening

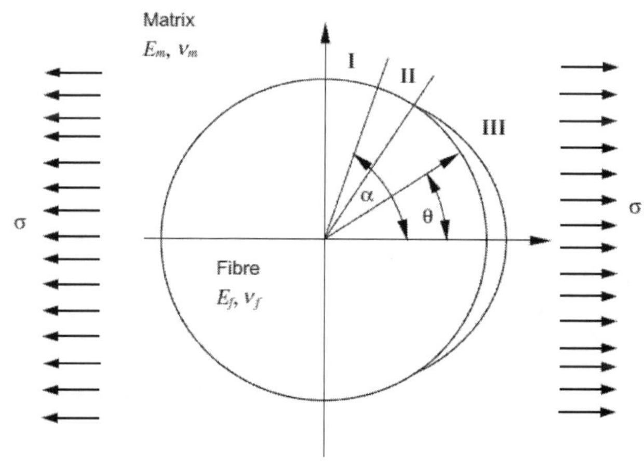


Figure 4: Schematic representation of the cross-section of the fibre embedded in matrix and loaded transverse to the fibre axis. The fibre/matrix interface is partially debonded and three regions at the interface can be distinguished: I, perfect bonding; II, debonded zone with debond face contact; III, open zone [11].

(overlapping) before oscillation of the solution may, depending on the debonding length, arise. The problem must be under these circumstances modelled in accordance with Comninou conclusions [13] as appears in Fig. 4 [11]. Region I represents perfect bonding between fibre and matrix whereas Regions II and III represents the debonded part, Region II corresponding to the contact zone and Region III corresponding to the opened part of the debonded zone. A physical explanation of the relative sizes of overlapping predicted by the analytical solutions and the BEM models is given in [10]. When the debonding starts to reach approximately 60°, a contact zone of physical meaning can be established.

## 5 Discussion about results

If the effect of the fibres in the laminate is taken into consideration (results labelled as "fibre in laminate"), the large stiffness of the specimen in the direction of the thickness causes that the crack behaves as in the plain strain model, and a constant *J*-integral value is obtained along the entire crack front. The two other sets of results are in accordance with the limiting cases for which the elastic properties of the isotropy plane are put the same to those of the matrix and the fibre. If the elastic properties are those of the matrix, the *J*-integral value is practically the same to that obtained for the single fibre example at the interior of the specimen, but it decreases at the free surface. Lastly, when the elastic properties of the isotropy plane are those of the fibre (the most rigid of all cases analyzed) *J*-integral presents its lowest level, and analogously to the homogenous case it introduces a constant value along the complete crack face.

The results facilitate the interpretation of experimental observations as those reported by Meurs [14] who tested a single glass-fibre-reinforced sample subject to lateral loading with four increasing steps when verified initiation of debonding.

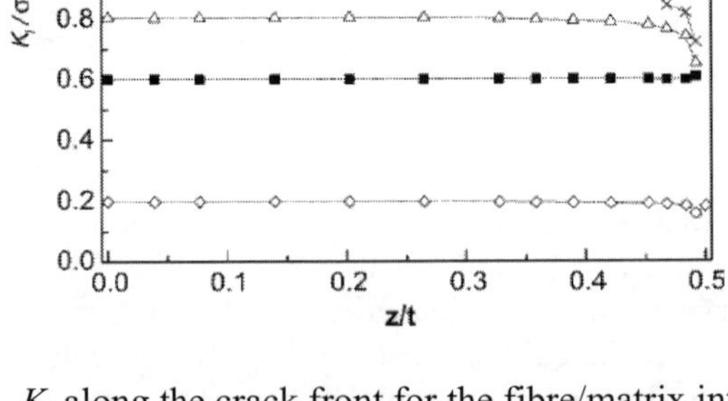


Figure 5: $K_I$ along the crack front for the fibre/matrix interface crack.

Note that in accordance with the numerical results for the single fibre example, debond crack initiates at the specimen surface where the maximum *J*-integral value is achieved.

Stress intensity factor issues are shown for the three modes of cracking eg in Fig. 5. Results are normalized with respect to $\sigma^{\infty}\sqrt{\pi a}$. Fig. 5 permits watching that the characteristics of $K_I$ values are obtained for the cases with the largest material mismatch between the fibre and the vicinity material, only with the exception of the limiting case for which the elastic properties of the isotropy plane are put on identical with those of the matrix, $K_I$ results present a constant value along the majority of the crack front.

## 6 Solution to biomaterial interface cracks [15]

Problems of fracture mechanics as far as cracks at the interface between dissimilar materials are of interesting owing to their relevance to issues of debonding or delamination in composite materials. Concurrently, analytical studies (eg [16, 17, 19]) have demonstrated that the stress field close to an interface crack tip shows an oscillatory singularity. Even though the oscillation region is restricted to the vicinity of the crack tip, it has become visible to put on a challenge to the FEM and BEM, in precisely predicting the individual SIF for interface crack problems. The primary advantage of the BEM over the FEM in solving fracture problems is that only the boundary of the domain is discretized. Substantial amount of multi-domain BEM studies has been devoted to problems involving bimaterial interface cracks. The multi-domain boundary contour method or MBCM technique cas also be applied to solve bimaterial interface fracture problems where the SIFs are among interestingly parameters.

In the case of bimaterial interface cracks, the complex SIFs, gained by dint of the displacement correlation technique, are related by (see [18])

$$K_1 + iK_2 = \lim_{r\to 0}\frac{2\cosh(\pi\varepsilon)}{C}\sqrt{\frac{2\pi}{r}}(1+2i\varepsilon)(\Delta u_n + i\Delta u_t)\left(\frac{r}{l}\right)^{-i\varepsilon} \tag{14}$$

where $\Delta u_t$ and $\Delta u_n$ are the tangential and normal components of the crack opening displacement, $l$ is a random length used to normalize the distance $r$ from the crack tip, and

$$C = \frac{\kappa_1+1}{G_1} + \frac{\kappa_2+1}{G_2}$$

$$\varepsilon = \frac{1}{2\pi}\ln\left|\left(\frac{\kappa_1}{G_1}+\frac{1}{G_2}\right)/\left(\frac{\kappa_2}{G_2}+\frac{1}{G_1}\right)\right|$$

$$\kappa_j = \begin{cases} 3-4\nu_j & \text{(plane strain)} \\ (3-\nu_j)/(1+\nu_j) & \text{(plane stress)} \end{cases} \tag{15}$$

with $G_1$ and $\nu_j$ as the shear modulus and Poisson´s ratio of the material $j$.

In the context of the MBCM, the above limit is at a rough estimate evaluated by using a small value of $r$ which is the distance from the current crack tip to a nearly node on the crack-tip element. Therefore, the COD components in eqn (14) are those at the selected node, determined in the form:

$$\begin{aligned} \Delta u_t &= \Delta u_1 \cos\theta + \Delta u_2 \sin\theta, \\ \Delta u_n &= -\Delta u_1 \sin\theta + \Delta u_2 \cos\theta \end{aligned} \tag{16}$$

where $\theta$ is the crack bearings at the said crack tip, $\Delta u_i = u_i^{upper} - u_i^{lower}$ where $i$ = 1,2, and $u_i^{upper}$ and $u_i^{lower}$ are the nodal displacement components of the upper and lower faces of the crack being specified from the MBCM analysis.

In general, it has been indicated that more accurate SIFS are obtained if the end-node of the crack-tip element is chosen for determining r instead of the mid-node (see Fig. 6). This may be clarified by the fact that if $r$ is too small, the numerical results are degenerated by the oscillatory stress field region near the crack tip.

For cracks in a homogenous material, $\varepsilon = 0$ and $C = 2\,(\kappa + 1)/\mu$ (according to eqn (15)). As a result, eqn (14) comes to be

$$K_{\mathrm{I}} + iK_{\mathrm{II}} = \frac{G}{\kappa+1}\sqrt{\frac{2\pi}{r}}\left(\Delta u_n + i\Delta u_t\right) \tag{17}$$

where the notations $K_{\mathrm{I}}$ and $K_{\mathrm{II}}$ are used to name the mode-I and mode-II. SIFs for homogeneous materials.

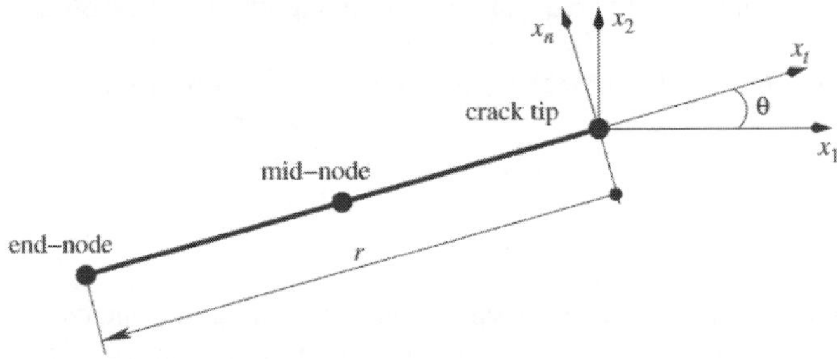


Figure 6: Crack-tip element.

## 7 Conclusion

Boundary element methodology for the three-dimensional analysis of bimaterial interface cracks has been presented. The interface crack study is spoken to, using a multidomain BEM formulation so as to be the reason for the different material properties at both sides of the crack. The parameters of fracture mechanics, namely $J$-integral and stress intensity factors, are calculated along the crack front employing the Energy Domain Integral and the $M_1$-integral methodologies.

These are implemented as a post-processing technique, and so it can be applied to the results from a particular model at a later stage. The carrying out exploits the efficiency of the boundary integral equation to directly obtain the demanded displacement derivatives, stress and strain fields from their boundary integral symbolizations.

The suggested numerical aid is employed to analyze the problem of a fibre/matrix interface crack under transverse loading in order to assess its three-dimensional nature. This three-dimensional model will in any case permit in the future the possibility of the interaction between longitudinal and circumferential growth of the crack to be studied. Obtained results show the leading role played by the relative elastic properties of the fibre, the matrix and the laminate on the state of mixed mode fracture. The instance of a single debonded fibre in a homogenous panel sets up the most severe condition, because it presents a large material mismatch with low lateral constraint. On the other hand, a fibre in a laminate behaves very similarly to a plain-strain case analysis, indicating that three-dimensional effects are very poor for this case. Lastly, the computed mixed-mode stress intensity factor indicates, that for most of the material combinations studied, $K_{\mathrm{I}}$ does not present important changes along the crack face, inclusive of the zone in the vicinity of the free lateral surface. By contrast, the boundary layer effect is noticeable for $K_{\mathrm{II}}$ and $K_{\mathrm{III}}$.

## Acknowledgement

The author gratefully acknowledges the financial support of the presented research by the Grant Agency of the Czech Republic (project No. 103/09/1580).

## References

[1] Cho, S.B., Lee, K.R., Coi, Y.S., Yuuki, R., Determination of stress intensity factors and boundary element analysis for interface cracks in dissimilar anisotropic materials. *Eng Fract Mech*, **43**, pp. 603-614, 1992.

[2] Yuuki, R., Xu, J-Q., Boundary element analysis of dissimilar materials and interface crack. *Comput Mech*, **14**, pp. 116-127, 1994.

[3] Cisilino, A.P., Ortiz, J.E. Three-dimensional boundary element assessment of a fibre/matrix interface crack under transverse loading. *Computers and Structures*, **83**, pp. 856-869, 2005.

[4] Tan, C.L., Gao, Y.L., Treatment of bimaterial interface crack problems using the boundary element method. *Eng Fract Mech*, **36**, pp. 919-932, 1990.

[5] Yuuki, R., Cho, S.B., Efficient boundary element analysis of stress intensity factors for interface cracks in dissimilar materials. *Eng Fract Mech*, **34**, pp. 179-188, 1989.

[6] Miyazaki, N., Ikeda, T., Soda, T., Munakata, T., Stress intensity factor analysis of interface crack using boundary element method-Application of contour-integral method. *Eng Fract Mech*, **45**, pp. 599-610, 1993.

[7] Chen FHK, Shield RT. Conservation laws in elasticity of the J-integral type. *J Appl Math Phys (ZAMP)*, **28**, pp.1-22, 1977.
[8] Gosz, M., Dolbow, J., Moran, B., Domain integral formulation for stress intensity factor computation along curved three-dimensional interface cracks. *Int J Solids Struct*, **35**, pp. 1763-1783, 1998.
[9] Nagashima, T., Omoto, Y., Tani, S., Stress intensity factor analysis of interface cracks using X-FEM. *Int J Num Meth Eng*, **56**, pp. 1151-1173, 2003.
[10] Cisilino, A.P., Aliabadi, M.H., BEM implementation of the energy domain integral for the elastoplastic analysis of 3D fracture problems. *Int J Fract*, **96**, pp. 229-245, 1999.
[11] Varna, J., París, F., del Caňo, J.C., The effect of crack-face contact on fibre/matrix debonding in transverse tensile loading. *Compos Sci Technol*, **57**, pp. 523-532, 1997.
[12] Toya, M.A., Crack along the interface of a circular inclusion embedded in an infinite solid. *J Mech Phys Solids*, **22**, pp. 325-348, 1975.
[13] Comninou, M., The interface crack. *ASME J Appl Mech*, **44**, pp. 631-636, 1977.
[14] Meurs, P., Characterization of Microphenomena in transversely loaded composite material. *PhD Thesis*. Technische Universiteit Eindhoven, 1998.
[15] Phan, A.-V., Mukherjee, S., The multi-domain boundary contour method for interface and dissimilar material problems. *Engn Anal with BE*, **33**, pp. 668-677, 2009.
[16] Erdogan F. Stress distribution in a nonhomogeneous elastic plane with cracks. *J Appl Mech*, **30**, pp. 232–236, 1963.
[17] Rice, J.R., Sih, G.C., Plane problems of cracks in dissimilar media. *J Appl Mech*, **32**, pp.418–23, 1965.
[18] Nagashima, T., Omoto, Y., Tani, S., Stress intensity factor analysis of interface cracks using X-FEM. *Int J Numer Methods Eng*, 56, pp. 1151–1173, 2003.
[19] Rice, J.R., Elastic fracture mechanics concepts for interfacial cracks. *J Appl Mech*, **55**, pp. 98–103, 1988.

# Section 3
# Dynamics and vibrations

# An efficient FEM-BEM procedure for the multi-frequency acoustical analysis of lined ducts containing passive components

R. Maréchal, E. Perrey-Debain, J.-M. Ville & B. Nennig
*Université Technologique de Compiègne, France*

## Abstract

This paper deals with strategies for computing efficiently the propagation of sound waves in ducts with acoustic lining at its walls. Although efficient, these treatments seem to have reached their limit and there is still a need for considering other passive techniques to reduce further the sound radiation at the duct exit. In most cases of practical interest, these added acoustics components can be modeled as acoustic cavities, which are connected to the duct and can be either purely reactive or dissipative. The assessment of the efficiency of such a system requires a precise knowledge of the acoustic field in the duct. Although standard Finite Element (FE) software could, in principle, be used for this purpose, a full FE model would be extremely demanding, particularly in the mid-frequency range and this can have a negative impact when, for instance, some efficient optimizations are needed. In the present work, we present a new numerical procedure that judiciously exploits the benefit of the FEM and the BEM approach. Firstly, a set of FE eigenmodes are computed in the cavity to produce a numerical impedance matrix connecting the pressure and the acoustic velocity on the duct wall interface. Then an integral representation for the acoustic pressure in the main duct is used. The presence of acoustic liners on the walls of the duct is taken into account via an appropriate modal decomposition of the Green's function. Typical applications involving Helmholtz resonators and Herschel-Quincke tubes are presented. We show that our algorithm allows a very fast and accurate computation of the scattering matrix of such a system with a numerical complexity that grows very mildly with the frequency.
*Keywords: impedance matrix, modal decomposition, lined duct, green's function, integral equations.*

 
www.witpress.com, ISSN 1743-355X (on-line)
doi:10.2495/BE100091

## 1 Introduction

In a large number of sound generating devices, sound waves of large amplitude are being set up inside a tube and some of the acoustic energy propagates in the duct before being radiated into the open. The practical applications of such systems range from noise transmission in vehicle exhaust systems, through ventilation and air conditioning ducts, to sound propagation in the ducted regions of turbofan aircraft engines.

The acoustic energy flow reduction techniques for the duct noise problem can be divided into two categories. Reactive techniques specifically aim to alter the duct impedance by reflecting back most of the incident acoustic wave. In most cases of practical interest, these acoustics components can be modeled as acoustic cavities, which are connected to the duct. Dissipative techniques specifically aim to absorb the sound field as it propagates down the duct. These most commonly take the form of a lining material (porous and/or perforate metal panel) placed on the walls of the duct. While these absorbent materials are known to be typically effective at relatively high frequencies, reactive techniques are best applied to low frequency noise problems. For these reasons, these techniques seem to have reached their limit when used on their own and there is still a need to consider the benefit of these two combined in order to reduce further the sound radiation at the duct exit. The assessment of the efficiency of such a system requires a precise knowledge of the acoustic field in the duct. Although standard Finite Element (FE) software could, in principle, be used for this purpose, a full FE model would be extremely demanding as the number of variables is expected to grow like $f^d$ ($f$ is the frequency and $d = 2, 3$ the dimension of the discretized domain). To make the matter worse, the FE method is known to suffer from pollution errors, which can be avoided at the price of a very high discretization level, particularly in the medium and high frequency range [1]. This can have a negative impact when some efficient optimizations (geometry and positions of these added components for instance) are needed.

In this work, we present a new numerical procedure that judiciously exploits the benefit of the FEM and the BEM approach. The idea relies mainly on the concept of an impedance matrix that connects the pressure to the acoustic normal velocity on the duct-cavity interface. This considerably reduces the number of variables as only the interface needs to be discretized. The technique is based on the use of integral equations for the acoustic pressure with appropriate Green's functions. The theory is presented in Sections 3 and 4. The last section shows practical applications involving Helmholtz resonators [2] and Herschel-Quincke tubes [3]. The benefit of the present approach is shown both in terms of CPU time and model reduction, when compared to standard FE models.

## 2 Problem statement

The problem under consideration is illustrated in Fig. 1. It consists of a two-dimensional lined main duct (domain $\Omega$) of height $h$ which is connected to a single cavity $\Omega_c$. The inlet and outlet pipes (regions I and II) are identical, each having rigid walls at its boundaries $\Gamma_w$. We wish to evaluate the scattering matrix (or S-matrix) of this acoustic system, that is given incident pressure waves $P_{\mathrm{I}}^+$ and $P_{\mathrm{II}}^-$, we compute the scattered waves $P_{\mathrm{I}}^-$ and $P_{\mathrm{II}}^+$. We call $\Gamma_1$ (resp. $\Gamma_2$) the lined wall of the main duct with impedance $Z_1$ (resp. $Z_2$).

In the main duct, the acoustic pressure $p$ satisfies the Helmholtz equation

$$\Delta p + k^2 p = 0. \tag{1}$$

On the lined walls, a local impedance condition is prescribed:

$$Z_i = \frac{1}{Y_i} = \frac{p}{\mathbf{v}\cdot\mathbf{n}}, \qquad i = 1, 2. \tag{2}$$

Note the impedance is generally function of the frequency and the acoustic normal velocity is proportional to the pressure gradient, i.e. $\partial_n p = \mathrm{i}\omega\rho\mathbf{v}\cdot\mathbf{n}$ where $\mathbf{n}$ is the outward unit normal. Here, we adopt the $\mathrm{e}^{-\mathrm{i}\omega t}$ -convention, $k = \omega/c$ is the wave number, $c$ the celerity, $\omega$ the angular frequency, $\rho$ the fluid density. The transmission conditions at the artificial boundaries $\Gamma_{\mathrm{I}}$ and $\Gamma_{\mathrm{II}}$ are given from the pressure wave field in the inlet and outlet pipes. This is expressed as the usual modal series

$$P_j^{\pm} = \sum_{m=0}^{\infty} A_{j,m}^{\pm}\psi_m^0(x)\mathrm{e}^{\pm\mathrm{i}\beta_m^0 z} \tag{3}$$

where $j$ =I or II. Here the pair $(\psi_m^0, \beta_m^0)$ defines the classical propagative (or evanescent) mode in the rigid pipe (the superscript 0 refers to the rigid wall case). Finally, we require that $p$ and its normal derivative (i.e. the normal velocity) to be continuous across the duct-cavity interface $\Gamma$.

## 3 Impedance matrix: general theory

To simplify the analysis, we will consider a cavity filled with air and acoustic losses can be taken into account via a locally reacting treatment on the cavity wall. Let

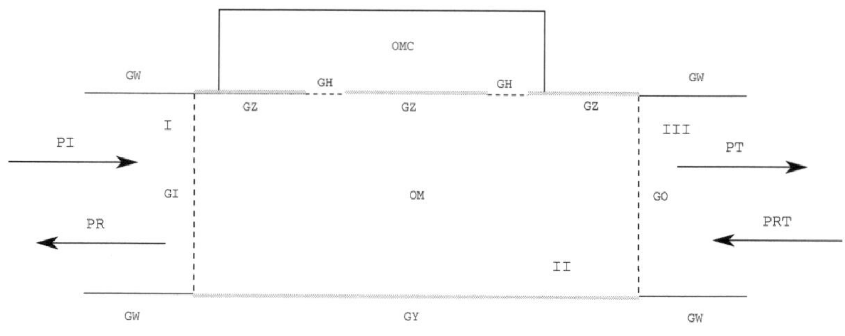


Figure 1: Main duct with a single passive component.

us first assume that the acoustic cavity $\Omega_c$ is closed, that is we enforce Neumann rigid boundary conditions on the apertures $\Gamma$. Standard results show there exists a complete set of orthogonal eigenfunction $\Phi_n$ belonging to the spectrum of the Laplacian operator (see [4])

$$\Delta\Phi_n = -k_n^2\Phi_n, \tag{4}$$

subject to the boundary conditions: $\partial_n\Phi_n = 0$ on the apertures $\Gamma$ and $\partial_n\Phi_n = \mathrm{i}\rho\omega Y_c\Phi_n$ everywhere else. Here, $Y_c$ denotes the complex wall admittance and $\partial_n$ denotes differentiation along the outward normal to the cavity. An important fact to keep in mind is that the boundary conditions depends on frequency (except in the limiting case of rigid walls) so that the set of eigenfunctions $\Phi_n$ and eigenvalues $k_n$ also are functions of $\omega$. Now, any given pressure field in $\Omega_c$ can be written as a weighted sum of the eigenfunctions. In particular, the Green's function for the cavity, which satisfies

$$\Delta G_c + k^2 G_c = -\delta(\boldsymbol{x} - \boldsymbol{y}), \tag{5}$$

is given by the infinite series

$$G_c(\boldsymbol{x}, \boldsymbol{y}, \omega) = \sum_{n=0}^{\infty} \frac{\Phi_n(\omega, \boldsymbol{x})\Phi_n(\omega, \boldsymbol{y})}{\omega^2 - \omega_n^2}, \tag{6}$$

where for the sake of clarity, we put $\omega_n = k_n(\omega)c$. Eigenfunctions are properly normalized so that application of the Green's theorem in the cavity yields the standard result

$$p(\boldsymbol{x}) = \int_\Gamma G_c\, \partial_n p \,\mathrm{d}\gamma(\boldsymbol{y}). \tag{7}$$

Note that this integral formulation is not only valid for any point $\boldsymbol{x}$ inside the cavity but also on the boundary $\Gamma$. Collocating (7) on $\Gamma$ leads, after discretization, to the linear system

$$\mathbf{p}_{\text{int}} = \mathbf{Z}(\omega)\, \tilde{\mathbf{F}}\, \partial_n \mathbf{p}_{\text{int}}. \tag{8}$$

Here, the vector $\mathbf{p}_{\text{int}}$ (resp. $\partial_n\mathbf{p}_{\text{int}}$) simply contains the discrete nodal values of the pressure (resp. normal derivative pressure) on the duct-cavity interface and, in this context, the frequency-dependent matrix $\mathbf{Z}$ can be interpreted as an impedance matrix. The matrix $\tilde{\mathbf{F}}$ stems from discretization of the boundary integral, its exact from is not essential for the moment but we will comment on this further on. Without loss of generality, it is easy to see from (6), that this matrix admits the following truncated eigenmode decomposition

$$\mathbf{Z}(\omega) = \left(\tilde{\boldsymbol{\Phi}}\, \mathbf{D}(\omega)\, \tilde{\boldsymbol{\Phi}}^{\mathrm{T}} + \mathbf{R}(\omega)\right). \tag{9}$$

The matrix $\tilde{\boldsymbol{\Phi}} = (\tilde{\boldsymbol{\Phi}}_1 \cdots \tilde{\boldsymbol{\Phi}}_N)$ contains in its columns the nodal value of the eigenfunctions $\Phi_n$ ($n = 0, 1, \ldots, N$) and $\mathbf{D}$ stands for the diagonal matrix with its diagonal entries: $(\omega_n^2 - \omega^2)^{-1}$. Here the tilde symbol means that we only retain the nodal values of these eigenmodes on the duct-cavity interface $\Gamma$. The interest for the decomposition (9) becomes clearer when the frequency of interest is taken well below the highest modal 'resonant' frequency (i.e. $|\omega - \omega_N|^{-1} \ll 1$). In this frequency range, the correction term $\mathbf{R}$ is generally weakly dependent on the frequency, so we can take the low order Taylor expansion

$$\mathbf{R}(\omega) \approx \mathbf{R}(\overline{\omega}) + (\omega - \overline{\omega})\frac{\partial \mathbf{R}}{\partial \omega}(\overline{\omega}) + \dots \tag{10}$$

where $\overline{\omega}$ is a reference value to be givenly precisely later. Once the matrix $\mathbf{R}(\overline{\omega})$ and its derivatives have been stored, the computation of (9) becomes a very fast and simple procedure. Its efficiency relies on the numerical computation of a sufficiently large set of eigenfunctions as well as finding the residual matrix. In this work, a finite element strategy is used to discretize the exact geometry of the cavity so that no assumption is made a priori regarding the shape of the acoustic eigenmodes. We consider a finite element mesh made of linear elements (two-dimensional triangles) and we denote by $\phi_i$ the associated piecewise linear shape function. Standard procedure leads to the following algebraic eigenmode problem:

$$\left(\mathbf{K} + \mathbf{C}(\omega) - \omega_n^2\mathbf{M}\right)\mathbf{\Phi}_n = 0 \tag{11}$$

where matrices involved are obtained from

$$(\mathbf{K})_{ij} = \int_{\Omega_c} \nabla\phi_i \nabla\phi_j \,\mathrm{d}\Omega \quad (\mathbf{M})_{ij} = \frac{1}{c^2}\int_{\Omega_c} \phi_i\phi_j \,\mathrm{d}\Omega, \tag{12}$$

and

$$(\mathbf{C}(\omega))_{ij} = -\mathrm{i}\rho\omega \int_{\partial\Omega_c \backslash \Gamma} Y_c(\omega)\phi_i\phi_j \,\mathrm{d}\gamma. \tag{13}$$

Equation (11) constitutes a generalized eigenvalue problem. Depending on the matrix size (i.e. the number of FE nodes in the whole cavity) and the number of modes retained in the series, the evaluation of the impedance matrix might be computationally expensive as this needs to be done for each frequency. This point will be discussed at the end of the section.

At this stage, we shall simplify the analysis a bit further by considering rigid boundary conditions only (i.e. we put $Y_c = 0$). In this case, the problem becomes linear in $\lambda = \omega^2$, i.e.

$$\mathbf{A}_0(\lambda_n^0)\mathbf{\Phi}_n^0 = 0 \quad \text{where} \quad \mathbf{A}_0(\lambda) = \mathbf{K} - \lambda\mathbf{M}. \tag{14}$$

Here the superscript 0 refers to the rigid wall case. A first set of mode ($n = 0, 1, \dots, N$) is then computed using appropriate large sparse eigenvalue problem solvers where it is understood that the truncation order $N$ is taken well below the

FE matrix size. Now, finding the residual matrix is a bit more tricky as we first need to go back to original direct problem:

$$\mathbf{A}_0(\lambda)\mathbf{p} = \mathbf{F}\,\partial_n \mathbf{p}_{\text{int}}, \tag{15}$$

where the vector $\mathbf{p}$ contains the value of the pressure at all nodes of the FE mesh. To ease the demonstration, we proceed to an appropriate elements reordering so that the rectangular matrix $\mathbf{F}$ looks like

$$\mathbf{F} = \begin{pmatrix} \tilde{\mathbf{F}} \\ \mathbf{0} \end{pmatrix} = \mathbf{I}_\Gamma \tilde{\mathbf{F}} \quad \text{with} \quad (\tilde{\mathbf{F}})_{ij} = \int_\Gamma \phi_i \phi_j \, \mathrm{d}\gamma. \tag{16}$$

Here $\mathbf{I}_\Gamma$ denotes the identity matrix for nodes on the interface only. Inverting (15) shows that the impedance matrix has the alternative form

$$\mathbf{Z}_0(\lambda) = \mathbf{I}_\Gamma^{\mathrm{T}}\, \mathbf{A}_0^{-1}(\lambda)\, \mathbf{I}_\Gamma \tag{17}$$

The residual matrix is then computed simply via

$$\mathbf{R}_0 = \mathbf{Z}_0 - \tilde{\mathbf{\Phi}}^0\, \mathbf{D}\, \left(\tilde{\mathbf{\Phi}}^0\right)^{\mathrm{T}} \tag{18}$$

and the first order derivative can be computed from

$$\frac{\partial \mathbf{R}_0}{\partial \lambda} = \mathbf{I}_\Gamma^{\mathrm{T}}\, \mathbf{A}_0^{-1} \mathbf{M}\, \mathbf{A}_0^{-1} \mathbf{I}_\Gamma - \tilde{\mathbf{\Phi}}^0 \frac{\partial \mathbf{D}}{\partial \lambda} \left(\tilde{\mathbf{\Phi}}^0\right)^{\mathrm{T}}. \tag{19}$$

Note that (i) the $\omega$-derivative is recovered from $\partial_\lambda \mathbf{R}_0 = 2\omega\partial_\omega \mathbf{R}_0$ and (ii) the $\lambda$-derivative of the diagonal matrix $\mathbf{D}$ is an easy task as eigenvalues $\lambda_n^0 = (\omega_n^0)^2$ are frequency-independent in the rigid wall case. Now exploiting the symmetry of $\mathbf{A}_0$, we see that $\mathbf{I}_\Gamma^{\mathrm{T}}\, \mathbf{A}_0^{-1} = (\mathbf{A}_0^{-1}\mathbf{I}_\Gamma)^{\mathrm{T}}$. Thus the important point to make is that the full inversion of $\mathbf{A}_0$ is not needed here. Only the first columns of $\mathbf{A}_0^{-1}$ corresponding to the interface nodes are active and these can be efficiently computed by solving successively $\mathbf{A}_0 \mathbf{v}_i = \mathbf{e}_i$ with appropriate sparse solvers ($\mathbf{e}_i$ is the column vector with zero elements with the unity on the $i$th line). It is clear that the reference value $\overline{\lambda}$ must be chosen away from the resonant value $\lambda_n^0$ to guarantee that the inversion of $\mathbf{A}_0$ is not spoiled by round off error. In this work we take $\overline{\lambda} = \lambda_1^0/2$ (the value $\overline{\lambda} = \lambda_0^0 = 0$ is proscribed here as it corresponds to the rigid motion resonance).

Now going back to the more general frequency dependent problem (11), the large size matrix system can be considerably reduced by searching the lossy eigenmode in the basis of the rigid modes, i.e.

$$\mathbf{\Phi}_n = \mathbf{\Phi}^0 \mathbf{q}_n. \tag{20}$$

Now by construction we have that

$$\left(\mathbf{\Phi}^0\right)^{\mathrm{T}} \mathbf{M}\mathbf{\Phi}^0 = \mathbf{I} \quad \text{and} \quad \left(\mathbf{\Phi}^0\right)^{\mathrm{T}} \mathbf{K}\mathbf{\Phi}^0 = \mathbf{\Lambda}^0, \tag{21}$$

where $\mathbf{\Lambda}^0$ is a diagonal matrix containing the rigid wall eigenvalues $\lambda_m^0$ $m = 0, 1, \ldots, N$ in ascending order. The eigenvalue problem becomes

$$\left(\mathbf{\Lambda}^0 + \left(\mathbf{\Phi}^0\right)^{\mathrm{T}} \mathbf{C}(\omega)\mathbf{\Phi}^0 - \omega_n^2\right) \mathbf{q}_n = 0. \tag{22}$$

Note that in most cases, high order modes are expected to behave almost like rigid wall modes (i.e. $\omega_n^2 \to (\omega_n^0)^2$ and $\mathbf{\Phi}_n \to \mathbf{\Phi}_n^0$ as $n$ increases) and the matrix $\mathbf{C}(\omega)$ can be treated as a perturbation term [4], this fact will be illustrated in the next section. Finally, the residual matrix is obtained following the same procedure from

$$\mathbf{Z}(\overline{\omega}) = \mathbf{I}_\Gamma^{\mathrm{T}}\, \mathbf{A}^{-1}(\overline{\omega})\, \mathbf{I}_\Gamma \quad \text{with} \quad \mathbf{A}(\omega) = \left(\mathbf{K} + \mathbf{C}(\omega) - \omega^2\mathbf{M}\right). \tag{23}$$

The first order derivative matrix is more difficult to obtain as all quantities involved are now frequency dependent and this will not be dealt with here.

## 4 S-matrix of the acoustic system

In the main duct, the theory starts by introducing the lined-walled duct Green's function satisfying the usual modal radiation condition on both ends of the main duct, i.e.

$$G(\boldsymbol{x}, \boldsymbol{x}_0) = \sum_{n=0}^{\infty} \frac{\psi_n(x)\psi_n(x_0)}{-2\mathrm{i}\beta_n}\, \mathrm{e}^{\mathrm{i}\beta_n|z-z_0|} \tag{24}$$

where $\boldsymbol{x} = (x, z)$ and $\boldsymbol{x}_0 = (x_0, z_0)$ are two points in the propagative domain $\Omega$. Function $\psi_n$ is the transverse mode, i.e.

$$\psi_n'' + \alpha_n^2 \psi_n = 0 \tag{25}$$

and satisfying the lined-wall conditions (2) on both sides $x = 0$ and $x = h$ (here the prime signifies derivation with respect to the transverse coordinate $x$). Again, such an eigenmode problem could be solved using FE techniques. However, an analytical approach is adopted here. It suffices to look for solutions of the form $\psi = A\cos(\alpha x) + B\sin(\alpha x)$. The transverse 'resonant' wavenumbers are found so that the $2 \times 2$ system

$$\mathsf{M}(\alpha) \begin{pmatrix} A \\ \alpha B \end{pmatrix} = 0 \tag{26}$$

admits non trivial solutions. It is more convenient for the analysis to search the axial wavenumber $\beta$, so we put $\alpha = \sqrt{k^2 - \beta^2}$ and solve the dispersion equation

$$g(\beta) = \det \mathsf{M}(\sqrt{k^2 - \beta^2}) = 0. \tag{27}$$

Using the fact that $g$ is analytic over the complex plane (in $\beta$), low order modes are numerically calculated with a technique described by Kravanja and Van Barel [5] and already used by the authors for the modeling of dissipative silencers [6].

Higher order modes are recovered using asymptotic expansions. This is easily achieved by expanding the transverse wavenumbers as the truncated series

$$\alpha_n h = \alpha_n^0 h + \sum_{k=1}^{K} \frac{a_k}{n^k}, \tag{28}$$

where $\alpha_n^0 = n\pi/h$ are the transverse wavenumbers of the rigid wall duct. This simple example reveals that lined wall modes behave almost like the rigid ones when $n$ is taken sufficiently large. In our applications, we observed that $n > 10$ was a sufficient condition for the series (28) to converge quickly with a relatively small number of terms ($K = 5$).

Using the Green's theorem, the pressure anywhere in the lined duct ($\Omega$ domain) is given via the integral representation

$$p = \int_{\Gamma_{\rm I} \cup \Gamma_{\rm II}} (G\, \partial_n p - p\, \partial_n G)\ \mathrm{d}\gamma + \int_{\Gamma} G\, (\partial_n p - \mathrm{i}\rho\omega Y_2 p)\ \mathrm{d}\gamma.$$

The discretization of this equation is carried out in two steps. First, collocating (29) at the FEM nodes of the duct-cavity interface leads to

$$\mathbf{K}_{\Gamma,\Gamma}\, \partial_n \mathbf{p}_{\rm int} + \mathbf{K}_{\Gamma,\rm I}\, \mathbf{A}_{\rm I}^- + \mathbf{K}_{\Gamma,\rm II}\, \mathbf{A}_{\rm II}^+ = \mathbf{F}_{\Gamma,\rm I}\, \mathbf{A}_{\rm I}^+ + \mathbf{F}_{\Gamma,\rm II}\, \mathbf{A}_{\rm II}^- \tag{29}$$

where vectors $\mathbf{A}_j^{\pm}$ contain the modes amplitudes $A_{j,m}^{\pm}$, ($j$=I,II). The first block matrix

$$\mathbf{K}_{\Gamma,\Gamma} = -\mathbf{Z}(\omega)\,\tilde{\mathbf{F}} - \mathbf{G} - \mathrm{i}\rho\omega Y_2 \mathbf{G}\mathbf{Z}(\omega)\,\tilde{\mathbf{F}} \tag{30}$$

stems from the self interaction of the acoustic pressure at the interface. Here the Green matrix $\mathbf{G}$ stems from the discretization of the second integral in (29). Other matrices are built by simply substituting $p = P_j^+ + P_j^-$ and $\partial_n p = \partial_n (P_j^+ + P_j^-)$ with ($j$=I,II) in the first integral of (29). This operation requires the computation of the coupling coefficients $C_{mn}$ given by the overlap integrals

$$C_{mn} = \int_0^h \psi_m^0(x)\psi_n(x)\ \mathrm{d}x, \tag{31}$$

which also arise in standard mode matching techniques [6]. The system (29) is completed by taking the evaluation point in the integral equation on the inlet and outlet boundaries. An additional set of equations is then produced by projecting (29) onto the hard-wall modes basis to give

$$\mathbf{K}_{j,\Gamma}\, \partial_n \mathbf{p}_{\rm int} + \mathbf{K}_{j,\rm I}\, \mathbf{A}_{\rm I}^- + \mathbf{K}_{j,\rm II}\, \mathbf{A}_{\rm II}^+ = \mathbf{F}_{j,\rm I}\, \mathbf{A}_{\rm I}^+ + \mathbf{F}_{j,\rm II}\, \mathbf{A}_{\rm II}^- \tag{32}$$

for both boundaries $j$=I and II. Finally, the scattered modes amplitudes $\mathbf{A}_{\rm I}^-$ and $\mathbf{A}_{\rm II}^+$ are found by solving (29) together with (32). In practice, the summation in (3) is limited to the number of propagative modes as well as some evanescent modes which are included to ensure a precise approximation of the pressure field in the inlet and outlet boundaries. Thus, the scattering matrix system is of a relatively small size in the present study.

## 5 Results and concluding remarks

The first example concerns the effect of a HQ tube placed in the lined section of the duct (see Fig. 2). The width of the main duct is $h = 2a = 0.04859$ m and the length of the liner is 0.6 m. The results are presented with respect to the dimensionless variable $ka$. The study is carried out from very low frequency up to $ka = 2.23$ (which corresponds to 5000 Hz) with a stepsize of 1 Hz. In the overall frequency range, the incident pressure is a plane wave. The first cut-off frequency occurs at $ka = \pi/2$. For the lined wall, the impedance value are chosen as to be in line with perforate plates encountered in the aeronautic industry. Thus, we take $Z_1 = \rho c(2 + 2\mathrm{i})$ and $Z_2 = \rho c(1 + 1\mathrm{i})$. Note that the expected frequency dependence of the impedance is not taken into account here but this can be easily included in the analysis.

Here, we are interested in the Transmission Loss (TL) is defined as the ratio of transmitted acoustic power with respect to the incident one, that is

$$\mathrm{TL} = -10\log_{10}\left(\frac{1}{\beta_0^0\left|A_{\mathrm{I},0}^+\right|^2}\sum_{m\geq 0}\beta_m^0\left|A_{\mathrm{II},m}^+\right|^2\right), \tag{33}$$

where the summation is limited to propagative modes only. In order to validate the method, a full finite element model is used. Radiation conditions at both ends of the duct have been implemented using the DtN map [7]. In Fig. 3 are plotted the TL calculated from our method and the FE model. The HQ tube is about 27 cm long with a width of 4.6 cm and the first 250 modes are included for the impedance matrix calculation (the FE mesh contains about 2,500 elements). The very good agreement validates the present method and the small discrepancies noticeable at high frequency are thought to be due to the FE model which starts loosing accuracy. The number of FE nodes is very large compared to number of variables used in our model (i.e. the number of nodes at the interface + the number of modes in the summation in (3)) which does not exceed 30, this is reported in Table 1. At higher frequency, this number is expected to grow very mildly with the frequency whereas the FE model would quickly become intractable because of the

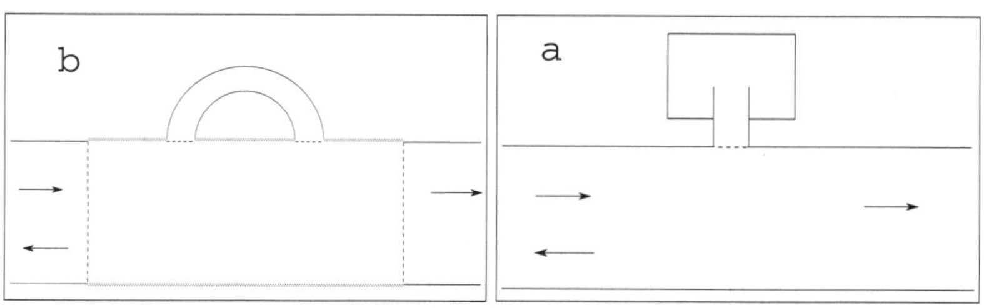


Figure 2: Typical acoustic components. (a): HQ tube connected to a lined section of a duct (gray color). (b): Helmholtz resonator connected to a rigid-wall duct.

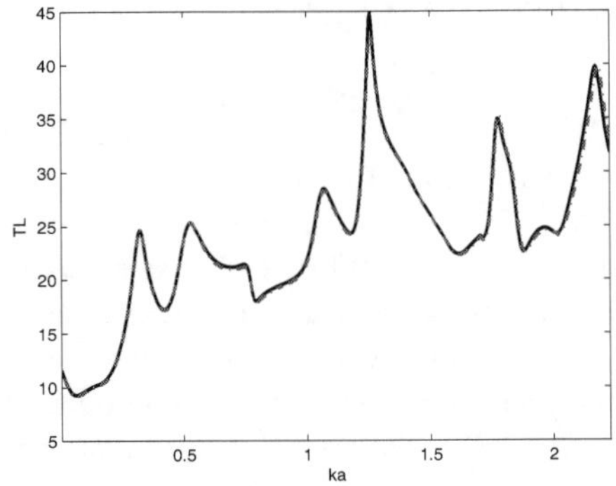


Figure 3: HQ tube in a lined duct : —— our method, - · - full FE model.

Table 1: Comparison of the proposed method with a full FE model. (*) time corresponding to 5,000 TL calculation on a single PC using MATLAB.

| | CPU time$^{(*)}$ | FE mesh | Total dof |
|---|---|---|---|
| HQ tube | 10 min | 2,500 | 30 |
| Helm. resonator | 30 s | 5,000 | 10 |
| full FE model | 75 min | 12,000 | 12,000 |

computational overhead. Note that (i) a large part of the CPU time is due to the calculation of the roots of the dispersion equation (27) and this could be largely reduced by using fastest methods such as the Newton-Raphson method and (ii) in all calculations, the modal series for the lined duct Green's function is truncated by keeping the first 250 terms in the series to ensure that the diagonal dominant coefficients of the interface-interface interaction matrix $\mathbf{K}_{\Gamma,\Gamma}$ are computed with sufficient accuracy.

The second example concerns the effect of one Helmholtz resonator connected to a rigid wall duct. This example is taken from [2] and shown in Fig. 2. For the sake of illustration, Fig. 4 shows typical eigenmode patterns for three resonators with different neck extensions (in order of appearance: straight, with conical contraction and with conical expansion). Again, a comparison with a full FE model gives perfect agreements as shown in Fig. 5. Here, the main duct is rigid and there is no need to find the lined duct Green's function, this explains the very low CPU time (30 s only). The right figure shows the effect of the extended necks on the TL.

Through these examples, the method presented in this work has shown to be extremely beneficial, both in term of CPU time and model reduction, when compared to standard FE models. Work is on going by the present authors to extend

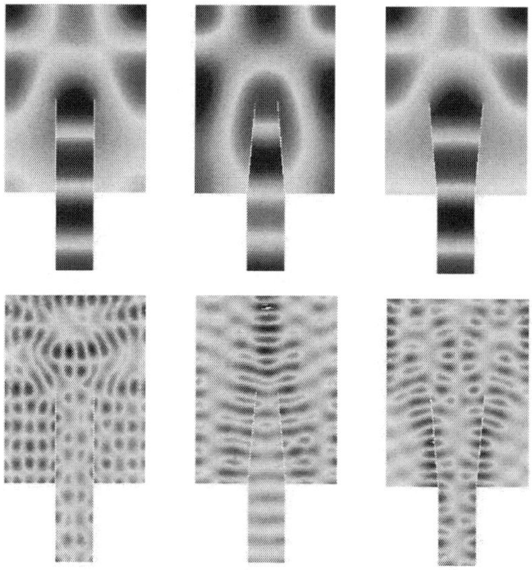

Figure 4: Eigenmodes in the Helmholtz resonators, top: Mode #13 (around 2,800 Hz) and bottom: Mode #215 (around 15,400 Hz).

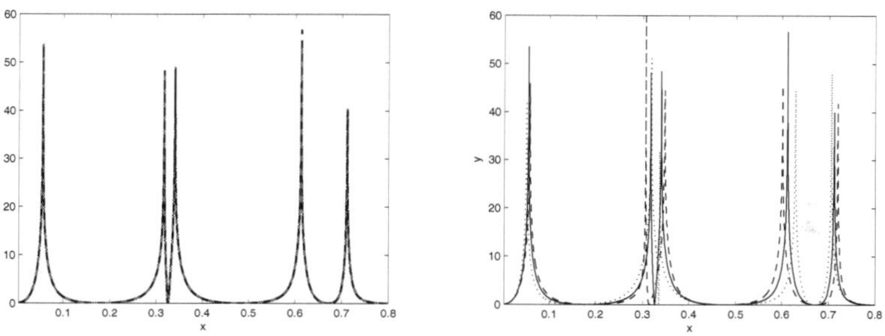


Figure 5: Left: Helmholtz resonator with straight extended neck connected to a rigid wall duct : —— our method, – · – full FE model. Right: Influence of the extended necks on the Transmission Loss; solid line: straight, dashed line: with conical expansion, dotted line: with conical contraction.

the method for more realistic 3D configurations. The presence of an absorbent material in the cavity will also be considered for future work. Finally, we think that the method proposed in this work could be extended for the analysis of other noise reduction techniques such as dissipative silencers for instance.

## References

[1] Gerdes, K. & Ihlenburg, F., On the pollution effect in FE solutions of the 3D-Helmholtz equation. *Comput Methods Appl Mech Engrg*, **170(1)**, pp. 155 – 172, 1999.

[2] Selamet, A. & Lee, I., Helmholtz resonator with extended neck. *J Acoust Soc Am*, **113(4)**, pp. 1975–1985, 2003.

[3] Selamet, A. & Dickey, N., The Herschel-Quincke tube: A theoretical, computational, and experimental investigation. *J Acoust Soc Am*, **96(5)**, pp. 3177–3185, 1994.

[4] Morse, P.M. & Ingard, K.U., *Theoretical Acoustics*. Princeton University Press, 1968.

[5] Kravanja, P. & Van Barel, M., *Computing the Zeros of Analytic Functions*. Springer, 2000.

[6] Nennig, B., Perrey-Debain, E. & Ben Tahar, M., Mode matching method for cylindrical dissipative silencers with poroelastic material. *The 9th Int Conf on Mathematical and Numerical Aspects of Waves*, 2009.

[7] Harari, I., Patlashenko, I. & Givoli, D., Dirichlet-to-neumann maps for unbounded wave guides. *J Comput Phys*, **143(1)**, pp. 200–223, 1998.

# Inverse scattering analysis for an elastic half space based on a fast volume integral equation method

T. Touhei, T. Takagishi, Y. Wajima & K. Kuranami
*Department of Civil Engineering, Tokyo University of Science, Japan*

## Abstract

A method for the inverse scattering analysis is presented for an elastic half space. The volume integral equation method is introduced here to reconstruct fluctuation of the medium from scattered waves observed at the free surface. The Born approximation is applied to the volume integral equation to formulate the equation for the inverse scattering analysis. The equation is solved by means of the Krylov subspace iteration method and the fast generalized Fourier transform. Several numerical calculations are carried out to investigate the solutions of the inverse scattering analysis.
*Keywords: inverse scattering problem, volume integral equation, elastic half space, fast generalized Fourier transform, Krylov subspace iteration method, Born approximation.*

## 1 Introduction

The analysis of scattered waves is an important issue in fields such as earthquake engineering, non-destructive testing and identification of energy resources. The integral equation methods have been efficient tools for both the forward and inverse scattering analysis. For example, Colton and Kress (1998) presented a survey of a vast number of articles on forward and inverse scattering analyses [1]. Among the integral equation methods, the volume integral equation known as the Lippmann-Schwinger equation has an advantage in that it presents a mathematical relationship between the fluctuation of the medium and scattered wave field. As a result, a number of examples of applications of the volume integral equation are increasing recently in spite of its deficiency, that is the requirement of a huge scale

WIT Transactions on Modelling and Simulation, Vol 50, © 2010 WIT Press
www.witpress.com, ISSN 1743-355X (on-line)
doi:10.2495/BE100101

and dense matrix for numerical analyses. Several methods for the volume integral equation presented recently are aimed to resolve the deficiency of the volume integral equation (for example, [2–4]). Even in this situation, application of the volume integral equation to inverse scattering analysis of an elastic half space is still a task for the future.

In this article, a method for inverse scattering analysis is presented for an elastic half space by means of the volume integral equation. The inverse scattering analysis here is to reconstruct fluctuation of the medium from the scattered waves observed at the free surface. The formulation is carried out by applying the Born approximation to the volume integral equation. As a result, the equation is obtained for the reconstruction of fluctuation of the medium from scattered wave field. The introduction of the fast generalized Fourier transform [5] and the Krylov subspace iteration method [6] enables us to solve the equation. Several numerical examples by means of the proposed method are presented in this article.

## 2 Theoretical formulation

### 2.1 Problem dealt in this article

Consider a scattering problem shown in Fig. 1, in which a point source is applied to a free surface of an elastic half space. The wave field has a localized fluctuation and scattered waves are caused due to an interaction between the incident waves and fluctuation of the medium. The problem of this article is to reconstruct the fluctuation by means of scattered waves. We assume that the back ground structure of the wave field is already provided for the analysis.

According to Fig. 1, the set of the points for grids to observe the scattered waves is denoted by $\Gamma_O$ and those to reconstruct the fluctuation is denoted by $\Gamma_R$. To set up $\Gamma_R$, the properties of the waves due to a point source is used here. As shown in later, the waves due to a point source propagating to downward direction are found

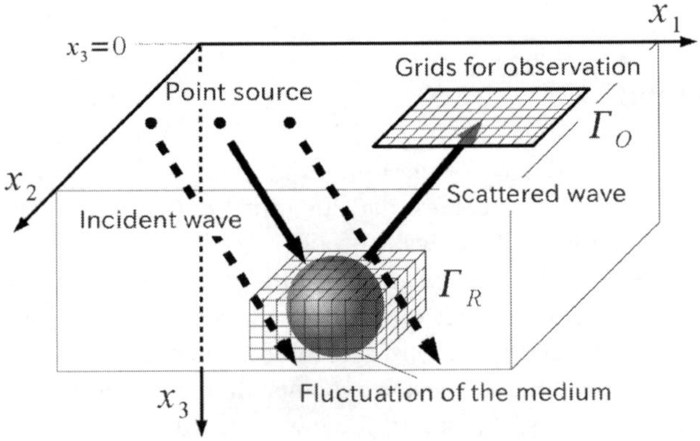


Figure 1: Concept of the scattering problem.

to exhibit a strong directionality. Therefore, by means of the position of the point source and the intensity of the scattered waves observed at the free surface, it is possible to set up grids for $\Gamma_R$. This concept is also described in Fig. 1.

The fluctuation of the medium is defined by the Lamé constants and the mass density such that

$$\begin{aligned} \lambda(x) &= \lambda_0 + \tilde{\lambda}(x) \\ \mu(x) &= \mu_0 + \tilde{\mu}(x) \\ \rho(x) &= \rho_0 + \tilde{\rho}(x), \quad (x \in \mathbb{R}^3_+) \end{aligned} \tag{1}$$

where $x \in \mathbb{R}^3_+$ is the spatial point, $\lambda_0$, $\mu_0$ and $\rho_0$ are the background constants of the Lamé and mass density and $\tilde{\lambda}$, $\tilde{\mu}$ and $\tilde{\rho}$ are their fluctuations. Note that the components of $x \in \mathbb{R}^3_+$ are expressed by

$$x = (x_1, x_2, x_3) \in \mathbb{R}^3_+ \tag{2}$$

where $x_3$ is the vertical coordinate with the positive direction downward and the free boundary is defined by $x_3 = 0$.

The inverse scattering analysis dealt in this article can now be stated as follows:

**Definition 1** *The inverse scattering analysis is to reconstruct the fluctuation of the medium $\tilde{\lambda}$, $\tilde{\mu}$ and $\tilde{\rho}$ at $\Gamma_R$ from information about the scattered waves observed at the grids $\Gamma_O$, the background structure of the medium and the incident wave field.*

## 2.2 Volume integral equation

The volume integral equation for the scattering problem shown in Fig. 1 is described as [4]

$$\begin{aligned} u_i(x) = &- \int_{\mathbb{R}^3_+} g_{ij}(x,y) N_{jk}(y) u_k^{(I)}(y, x_s) dy \\ &- \int_{\mathbb{R}^3_+} g_{ij}(x,y) N_{jk}(y) u_k(y) dy, \quad (x \in \mathbb{R}^3_+) \end{aligned} \tag{3}$$

where $u_i$ is the scattered wave field, $g_{ij}$ is the Green's function for an elastic half space due to the background structure, $N_{ij}$ is the differential operator constructed by the fluctuations of the medium, $u_i^{(I)}$ is the incident wave field caused by a point source and $x_s$ is the spatial point where the point source is applied. The subscript indices $i$, $j$ and $k$ for Eq. (3) denotes the components of the vectors or tensors to which the summation convention is applied.

The Green's function for an elastic half space used for the integral equation is defined by

$$\left((\lambda_0 + \mu_0)\partial_i\partial_j + \delta_{ij}\mu_0\partial_l\partial_l + \delta_{ij}\rho_0\omega^2\right) g_{jk}(x,y) = -\delta_{ik}\delta(x-y) \tag{4}$$

$$P_{ij}g_{jk}(x,y)=0 \text{ at } x_3=0 \tag{5}$$

where $\partial$ is the partial differential operator, $\delta_{ij}$ is the Kronecker delta, $\omega$ is the circular frequency, and $P_{ij}$ is the differential operator expressed as

$$[P_{ij}]=\begin{bmatrix}\mu_0\partial_3 & 0 & \mu_0\partial_1\\ 0 & \mu_0\partial_3 & \mu_0\partial_2\\ \lambda_0\partial_1 & \lambda_0\partial_2 & (\lambda_0+2\mu_0)\partial_3\end{bmatrix} \tag{6}$$

For numerical analysis here, the Green's function defined by Eqs. (4) and (5) is constructed by the fast generalized Fourier transform. By means of the Green's function, the incident wave field is calculated such that

$$u_k^{(I)}(x,x_s)=g_{kl}(x,x_s)\beta_l \tag{7}$$

where $\beta_l$ is the force vector of the point source. In addition, the explicit form of the differential operator $N_{ij}$ in Eq. (3) is expressed by

$$\begin{aligned}N_{ij}(x)=&-\left(\tilde{\lambda}(x)+\tilde{\mu}(x)\right)\partial_i\partial_j-\delta_{ij}\tilde{\mu}(x)\partial_k\partial_k\\ &-\partial_i\tilde{\lambda}(x)\partial_j-\delta_{ij}\partial_k\tilde{\mu}(x)\partial_k-\partial_j\tilde{\mu}(x)\partial_i-\delta_{ij}\tilde{\rho}(x)\omega^2\end{aligned} \tag{8}$$

### 2.3 Method for the inverse scattering analysis

At this stage, let us modify the volume integral equation into that represents the relationship between the observed scattered waves at $\Gamma_O$ and fluctuation of the medium at $\Gamma_R$. Application of the Born approximation to the volume integral equation shown in Eq. (3) yields

$$u_i(x)=-\int_{\mathbb{R}^3_+}g_{ij}(x,y)N_{jk}(y)u_k^{(I)}(y,x_s)dy \tag{9}$$

The right side of Eq. (9) can be modified so that the unknown quantities are the fluctuation of the medium. Let $q_k$ be the component of the states vector with respect to the fluctuation of the medium such that

$$(q_k)=\begin{pmatrix}\tilde{\lambda} & \tilde{\mu} & \tilde{\rho}\end{pmatrix}^T \tag{10}$$

Then, Eq. (9) becomes as follows:

$$u_i(x)=\int_{\mathbb{R}^3_+}g_{ij}(x,y)M_{jk}(y,x_s)q_k(y)dy \tag{11}$$

where $M_{ij}$ is the operator including the effects of the incident wave field, which is defined by

$$N_{jk}(y)u_k^{(I)}(y,x_s)=-M_{jk}(y,x_s)q_k(y) \tag{12}$$

The explicit form of $M_{jk}$ is expressed as

$$[M_{jk}] = \begin{bmatrix} \partial_1 E_{ll} + E_{ll}\partial_1 & \partial_1 E_{ll} + \eta_1 + 2E_{1l}\partial_l & \omega^2 u_1^{(I)} \\ \partial_2 E_{ll} + E_{ll}\partial_2 & \partial_2 E_{ll} + \eta_2 + 2E_{2l}\partial_l & \omega^2 u_2^{(I)} \\ \partial_3 E_{ll} + E_{ll}\partial_3 & \partial_3 E_{ll} + \eta_3 + 2E_{3l}\partial_l & \omega^2 u_3^{(I)} \end{bmatrix} \tag{13}$$

where $E$ and $\eta$ in Eq. (13) are defined by

$$\begin{aligned} E_{ij} &= (1/2)\left(\partial_i u_j^{(I)} + \partial_j u_i^{(I)}\right) \\ E_{ll} &= E_{11} + E_{22} + E_{33} \\ \eta_j &= \left(\partial_1^2 + \partial_2^2 + \partial_3^2\right) u_j^{(I)} \end{aligned} \tag{14}$$

Note that the summation convention is applied to the subscript index $l$ in Eq. (13).

At this stage, the generalized Fourier and its inverse transforms are applied to Eq. (11). In addition, assume that

$$q_k(y) = 0, \quad (y \in \mathbb{R}_+^3 \setminus \Gamma_R) \tag{15}$$

$$x \in \Gamma_O \tag{16}$$

for Eq. (11). Then, Eq. (11) becomes the equation for the inverse scattering analysis, which is in the following form:

$$u_i(x) = \mathscr{U}_{ij}^{-1} \hat{h}(\xi) \mathscr{U}_{jk} M_{kl}(y, x_s) q_l(y), \quad (x \in \Gamma_O), \tag{17}$$

where $\mathscr{U}$ and $\mathscr{U}^{-1}$ are the operators for the generalized Fourier and its inverse transforms, respectively, and $\hat{h}$ is the function related to the generalized Fourier transform of the Green's function expressed by

$$\hat{h}(\xi) = \frac{1}{\mu_0 \xi_3^2 - \rho_0 \omega^2 + i\epsilon} \tag{18}$$

where $\epsilon$ is an infinitesimally small positive number and $\xi$ is the wavenumber vector having components

$$\xi = (\xi_1, \xi_2, \xi_3) \in \sigma_p \cup \sigma_c \subset \mathbb{R}_+^3 \tag{19}$$

Note that $\sigma_p$ and $\sigma_c$ are the set of the wavenumber for the Rayleigh and the body waves, respectively. The detail of the discussions for these sets are given in the article [4].

Now, we can solve Eq. (17) by means of the fast generalized Fourier transform and the Krylov subspace iteration method. Note that the present method does not require the derivation of a coefficient matrix. During the iterative process, only the discretized transforms are repeatedly used for the analysis.

# 3 Numerical examples

## 3.1 Analyzed target model and incident wave

Preliminary to carry out an inverse scattering analysis, let us set up a target model for the analysis. The back ground Lamé constants and mass density for the model are $\lambda_0$=4.0 GPa, $\mu_0$=2.0 GPa and $\rho_0$=2.0 g/cm$^3$, respectively. Namely, the P wave velocity for the back ground structure of the wave field is 2 km/s and the S wave velocity for that is 1 km/s. The fluctuation of the medium defined by $\tilde{\lambda}$ and $\tilde{\mu}$ is shown in Fig. 2(a). The fluctuation of the Lamé constants are in the form of a cube whose amplitudes are 0.2 GPa. In addition, the fluctuated area is embedded at the depth of 3.0 km from the free surface.

An incident wave field used for the analysis is shown in Fig. 2(b), that is constructed by the fast generalized Fourier transform. For the discretization of the generalized Fourier transform, the intervals of the grids in the space domain and the number of grids for the $j$-th coordinate ($j = 1, 2, 3$) are set by $\Delta x_j = 0.25$ km and $N_j = 256$, respectively. The amplitude of the point source for the incident wave field is $1.0 \times 10^{10}$ N, whose direction is in vertical, and the excitation frequency is 1.0 Hz. As can be seen in Fig. 2(b), the high displacement amplitudes areas can be seen along the free surface as well as downward directions from the free surface. These areas are corresponding to the Rayleigh and the body waves, respectively. The body waves propagating to downward direction are found to show the strong directionality and expected to interact with the fluctuation of the medium.

Figures 3(a) and (b) show the propagation of scattered waves at the free surface and the amplitude of the scattered waves in $x_1 - x_3$ plane, respectively. According to Fig. 3(a), the displacement of the scattered waves are larger in the forward region of the fluctuation of the medium. In addition, some spots showing larger

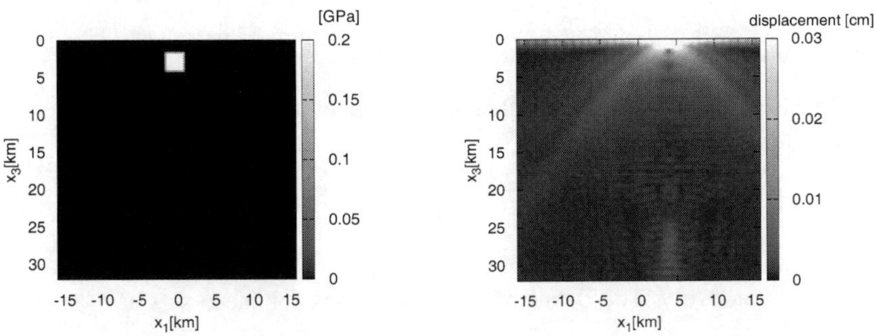


(a) fluctuations of $\tilde{\lambda}$ and $\tilde{\mu}$ (b) incident wave field

Figure 2: Fluctuation of the medium and incident wave field.

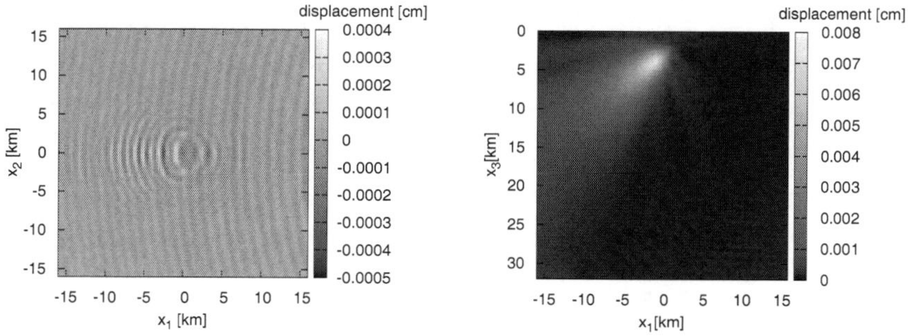


(a) displacement at the free surface

(b) displacement amplitudes in the vertical plane

Figure 3: The result of forward scattering analysis. The displacement of the scattered waves are shown.

displacement can be seen in the forward region of the fluctuation. These spots are caused due to the reflection of the body waves from the fluctuation of the medium, from which the directionality of the body waves can be recognized. It is also found from Fig. 3(b) that the outstanding of the displacement amplitudes in the forward direction also can be seen. The displacement amplitudes in the vertical plane is found to be much larger than that in the free surface. This indicates that the refection of the body waves from the fluctuated area to the free surface is not very large when compared to the scattered waves propagating to downward direction from the fluctuation of the medium.

### 3.2 Examples of the inverse scattering analysis

At this stage, let us reconstruct the fluctuation of the medium. Figures 4(a) and (b) show a range of data sampling of scattered waves at the free surface and the results of the reconstruction of $\tilde{\mu}$ at the depth of 3 km from the surface. Likewise, Figs 5(a) and (b) also show a range of data sampling and the results of the reconstruction of $\tilde{\mu}$ at the depth of 3 km. The differences of the data samplings between Figs. 4 and 5 are that Fig. 4 takes the data of scattered waves of the forward direction while Fig. 5 takes those of the backward direction.

It is found from Fig. 4(b) that the highest amplitude of the estimated fluctuation is 2.0 GPa. Since the target of the fluctuation was 0.2 GPa as shown in Fig. 2(a), the estimation of the fluctuation is higher here. According to the estimation shown in Fig. 5(b), the highest amplitude is 5.0 GPa. In this case, the estimated fluctuation is much higher than that shown in Fig. 4. Namely, the discrepancy between the target and estimation of the fluctuation increases in Fig. 5. The reason for this is clearly due to the properties of the sampled scattered waves, which convey the information for the fluctuation of the medium. These results indicate that sampling

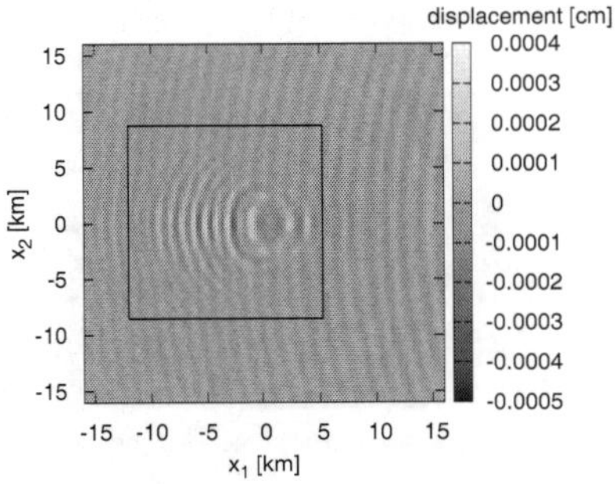


(a) range of the data sampling

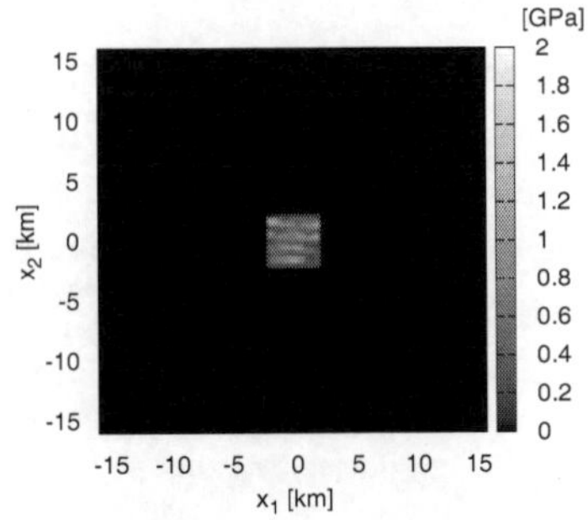


(b) reconstruction of $\tilde{\mu}$ at the depth of 3 km

Figure 4: The result of inverse scattering analysis.

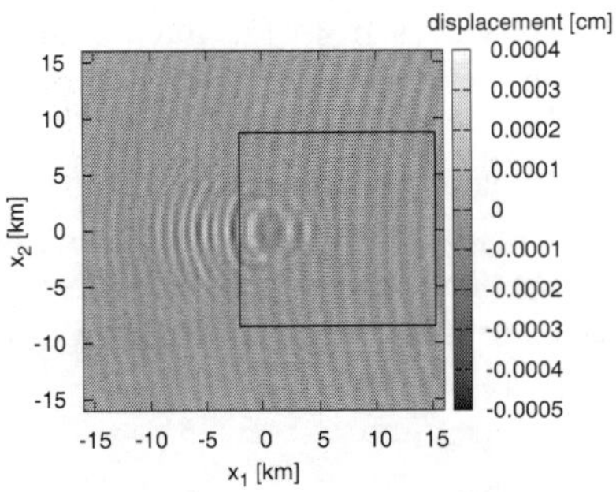


(a) range of the data sampling

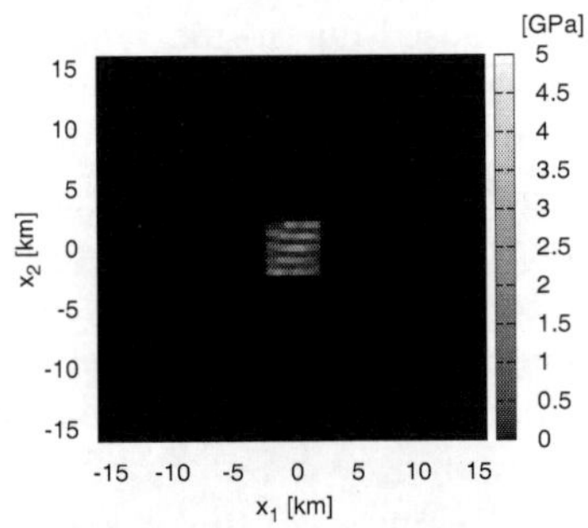


(b) reconstruction of $\tilde{\mu}$ at the depth of 3 km

Figure 5: The result of inverse scattering analysis.

of the scattered waves should be in the forward direction of the fluctuation. Note that the accuracy of the reconstructed fluctuation here has to be improved for the future. We have to examine the effects of the excitation frequency, the intervals of the grids, number of data sampling and among others.

The numerical calculations were carried out by a computer with an AMD Opteron 2.4 GHz processor. The CPU time needed for the present example based on the Bi-CGSTAB method was around 150 min, in which the number of the iterations was 10 times to obtain the results.

## 4 Conclusion

This article presented a method for inverse scattering analysis based on the volume integral equation method. The Born approximation was applied to the volume integral equation to obtain the equation for the inverse scattering analysis, which connected the scattered waves observed at the free surface and fluctuation of the medium. The equation was solved by the fast generalized Fourier transform and the Krylov subspace iteration method. Several numerical calculations were carried out to examine the solutions of the inverse scattering analysis. According to the numerical calculations, the amplitude of the estimated fluctuations of the medium were higher than that of the target. Properties of the sampled scattered waves were found to affect significantly to the results of the reconstruction of the fluctuation. Improvements of the accuracy of the results are the task for the future.

## References


[1] Colton, D. and Kress, R. (1998). *Inverse acoustic and electromagnetic scattering theory*, Berlin, Springer.

[2] De Zaeytijd,J., Bogaert,I. and Franchois, A. (2008). An efficient hybrid MLFMA-FFT solver for the volume integral equation in case of sparse 3D inhomogeneous dielectric scatterers, *Journal of Computational Physics*, **227**, 7052-7068.

[3] Yang, J., Abubaker, A., van den Berg, P.M., Habashy, T.M. and Reitich, F. (2008). A CG-FFT approach to the solution of a stress-velocity formulation of three-dimensional elastic scattering problems, *Journal of Computational physics*, **227**, 10018-10039.

[4] Touhei, T. (2009). Generalized Fourier transform and its application to the volume integral equation for elastic wave propagation in a half space, *International Journal of Solids and Structures*, **46**,52-73.

[5] Touhei, T., Takagishi, T. and Wajima, Y. (2009), Analysis of scattered elastic waves in a half space by means of the volume integral equation method. *Journal of applied mechanics,(JSCE)*, **12**, 1-8, in Japanese.

[6] Barrett, M., Berry, M., Chan, T.F., Demmel, J., Donato, J. M., Dongarra, J., Eijkhout, V., Pozo, R., Romine, C. and Van der Vorst, H. (1994). *Templates for the solution of Linear Systems: Building Blocks for Iterative Methods*, SIAM.

# Section 4
# Electrical engineering and electromagnetics

# Boundary element modeling of complex grounding systems: study on current distribution

D. Poljak[1], D. Ćerdic[2], V. Doric[1], A. Peratta[2],
V. Roje[1] & C. A. Brebbia[2]
[1]*University of Split, Croatia*
[2]*Wessex Institute of Technology, UK*

## Abstract

The assessment of current distribution induced along complex grounding systems has been undertaken using the corresponding antenna model and the set of Pocklington integro-differential equations for curved wires. The set of Pocklington equations is numerically handled via the Galerkin-Bubnov scheme of the Indirect Boundary Element Method (GB-IBEM) featuring the isoparametric elements. Some illustrative numerical results for the current distribution are presented in the paper.
*Keywords: boundary elements, antenna theory, set of Pocklington equations, grounding systems.*

## 1 Introduction

The analysis of complex grounding systems being important component in lightning protection systems (LPS) is of great interest in electromagnetic compatibility (EMC) and high voltage (HV) engineering. In general, grounding systems can be analyzed using the transmission line model (TLM) [1, 2], which principle feature is simplicity, or the antenna model (AM) [3–5], which is considered to be the rigorous approach, (but at the same time much more demanding, regarding both the formulation and numerical solution. The key-parameter in the antenna model is the equivalent current distribution flowing along grounding structures. Given the knowledge of the current distribution along the grounding system it is possible to determine other parameters of

WIT Transactions on Modelling and Simulation, Vol 50, © 2010 WIT Press
www.witpress.com, ISSN 1743-355X (on-line)
doi:10.2495/BE100111

interest, i.e. the scattered voltage, the input impedance and the transient impedance.

This paper represents an extension of the previous work of the authors on the subject which was related to the BEM analysis of a single grounding electrode [6, 7]. The antenna model of the grounding systems composed from straight conductors is based on the set of Pocklington's integro-differential equations for curved wires. The effect of a lossy ground is taken into account via the corresponding reflection coefficient [5] thus avoiding the rigorous approach based on the analytically demanding and numerically time consuming Sommerfeld integrals [8]. The set of Pocklington equations is numerically treated by means of the Galerkin-Bubnov scheme of the Boundary Element Method (GB-IBEM) with isoparametric elements [9]. Some illustrative numerical results for the current distribution along different grounding geometries are presented.

## 2 Wire grid model of the grounding system

The geometry of interest, shown in Fig 1, is the complex grounding system composed from interconnected conductors.

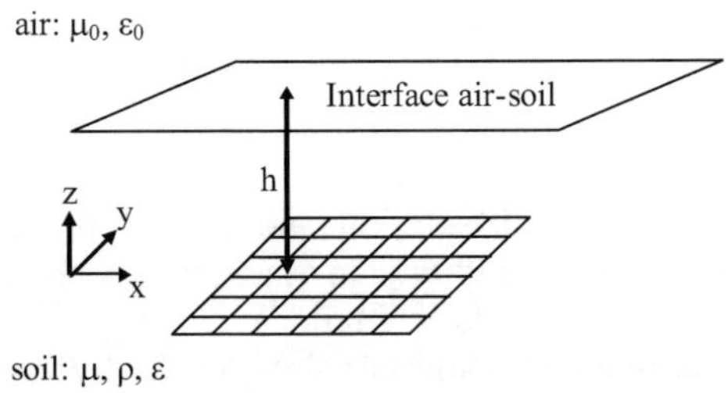


Figure 1: Geometry of a complex grounding system.

The set of Pocklington equiations for a configuration of interconnected buried wires can be obtained as an extension of the Pocklington equation for a single buried wire of arbitrary shape [7].

The Pocklington equation for a single grounding electrode can be derived by enforcing the continuity conditions for the tangential components of the electric field along the perfectly conducting (PEC) wire surface. An extension to the case of imperfectly conducting wire is straightforward [6, 7]. For the PEC wire the total field composed from the excitation field $\vec{E}^{exc}$ and scattered field $\vec{E}^{sct}$ vanishes:

$$\vec{e}_x \cdot \left( \vec{E}^{exc} + \vec{E}^{sct} \right) = 0 \quad \text{on the wire surface} \tag{1}$$

Starting from Maxwell's equations and Lorentz gauge the scattered electric field can be expressed in terms of the vector potential $\vec{A}$ :

$$\vec{E}^{sct} = -j\omega\vec{A} + \frac{1}{j\omega\mu\varepsilon}\nabla(\nabla\vec{A}) \tag{2}$$

which is defined by the particular integral:

$$\vec{A}(s) = \frac{\mu}{4\pi}\int_C I(s')g(s,s',s^*)\vec{s}\,'ds' \tag{3}$$

where $I(s')$ is the induced current along the wire, and $g(s,s')$ is the corresponding Green's function of the form:

$$g(s,s',s^*) = g_0(s,s') - \Gamma_{ref}\, g_i(s,s^*) \tag{4}$$

where $g_0(s,s`)$ denotes the lossy medium Green function, while $g_i(s,s`)$ arises from the image theory. These functions are given by:

$$g_0(s,s') = \frac{e^{-\gamma R_1}}{R_1} \qquad g_i(s,s') = \frac{e^{-\gamma R_2}}{R_2} \tag{5}$$

and $R_1$ and $R_2$ are the distances from the source point to the observation point, respectively, while the propagation constant of the lower medium is defined as:

$$\gamma = \sqrt{j\omega\mu\sigma - \omega^2\mu\varepsilon_{rg}\varepsilon_0} \tag{6}$$

where $\varepsilon_{rg}$ and $\sigma$ are the corresponding ground permittivity and conductivity, respectively.

The influence of a ground-air interface is taken into account via the reflection coefficient of the form [6]:

$$\Gamma_{ref} = \frac{\frac{1}{\underline{n}}\cos\theta - \sqrt{\frac{1}{\underline{n}} - \sin^2\theta}}{\frac{1}{\underline{n}}\cos\theta + \sqrt{\frac{1}{\underline{n}} - \sin^2\theta}}; \quad \theta = arctg\frac{|x-x'|}{2d}; \quad \underline{n} = \frac{\varepsilon_{eff}}{\varepsilon_0} \tag{7}$$

The complex permittivity of the lossy ground $\varepsilon_{eff}$ is given by:

$$\varepsilon_{eff} = \varepsilon_r\varepsilon_0 - j\frac{\sigma}{\omega} \tag{8}$$

Combining equations (1)–(3) and performing some mathematical manipulation leads to the integro-differential equation of the Pocklington type for the unknown current distribution induced along the thin wire of an arbitrary shape buried in a lossy half-space [9]:

$$\begin{aligned} E^{exc}(s) = &-\frac{1}{j4\pi\omega\varepsilon_{eff}}\int_{C'} I(s')\cdot\vec{s}\cdot\vec{s}\,'\cdot\left[\frac{\partial^2}{\partial s\partial s'} - \gamma^2\right]g_0(s,s')ds' + \\ &+\Gamma_{ref}\int_{C'} I(s')\cdot\vec{s}\cdot\vec{s}^{\,*}\cdot\left[\frac{\partial^2}{\partial s\partial s^*} - \gamma^2\right]g_i(s,s^*)ds' \end{aligned} \tag{9}$$

An extension to the case of multiple wires buried in a lossy medium is straightforward and results to the set of coupled Pocklington equations:

$$E_m^{exc}(s) = -\frac{1}{j4\pi\omega\varepsilon_{eff}} \sum_{n=1}^{M} \left[ \begin{array}{l} \int_{C'} I_n(s_n') \cdot \vec{s}_n \cdot \vec{s}_n' \cdot \left[ \frac{\partial^2}{\partial s_m \partial s_n'} - \gamma^2 \right] g_{0,nm}(s_m, s_n') ds' + \\ +\Gamma_{ref} \int_{C'} I_n(s') \cdot \vec{s}_m \cdot \vec{s}_n^* \cdot \left[ \frac{\partial^2}{\partial s_m \partial s_n^*} - \gamma^2 \right] g_{i,nm}(s_m, s_n^*) ds' \end{array} \right], \quad m = 1, 2, \ldots M \tag{10}$$

where $I_n(s')$ is unknown current distribution along the $n$-th wire, $E_m^{exc}(s)$ the excitation function on the $m$-th wire.

Finally, the grounding system is energized by the injection of the current source with one terminal connected to the grounding structure and the other one grounded at infinity [4]. Consequently, the left hand side of the equation (10) vanishes reducing the corresponding Pocklington equation to the homogenous one.

$$-\frac{1}{j4\pi\omega\varepsilon_{eff}} \sum_{n=1}^{M} \left[ \begin{array}{l} \int_{C'} I_n(s_n') \cdot \vec{s}_n \cdot \vec{s}_n' \cdot \left[ \frac{\partial^2}{\partial s_m \partial s_n'} - \gamma^2 \right] g_{0,nm}(s_n, s_n') ds' + \\ +\Gamma_{ref} \int_{C'} I_n(s') \cdot \vec{s}_m \cdot \vec{s}_n^* \cdot \left[ \frac{\partial^2}{\partial s_m \partial s_n^*} - \gamma^2 \right] g_{i,nm}(s_m, s_n^*) ds' \end{array} \right] = 0, \quad m = 1, 2, \ldots M \tag{11}$$

. This current source excitation is incorporated into formulation through the boundary conditions.

## 3 Boundary element procedures

The set of Pocklington integro-differential equations (11) is numerically handled by means of the Galerkin-Bubnov variant of Indirect Boundary Element Method (GB-IBEM). The Boundary Element solution technique used in this work is an extension of the method applied to single wire cases reported elsewhere, e.g. in [6, 7, 9].

The unknown current along the $n$-th wire segment is expressed by the sum of a linearly independent basis functions $f_{ni}$, with unknown complex coefficients $I_{ni}$:

$$I_n^e(s') = \sum_{i=1}^{n} I_{ni} f_{ni}(s') = \{f\}_n^T \{I\}_n \tag{12}$$

and the use of isoparametric elements yields:

$$I_n^e(\zeta) = \sum_{i=1}^{n} I_{ni} f_{ni}(\zeta) = \{f\}_n^T \{I\}_n \tag{13}$$

where $N_e$ is the number of local nodes on the element.

A linear approximation over a segment along $n$-the wire is used in this work as this choice was proved to be optimal in modeling of various wire structures [9]. The corresponding shape functions are given by:

$$f_i = \frac{1-\zeta}{2} \qquad f_{i+1} = \frac{1+\zeta}{2} \tag{14}$$

Applying the weighted residual approach with Galekin-Bubnov procedure (same test and basis functions) after performing a certain mathematical manipulations, the following matrix equation is obtained the set of Pocklington equations is transformed into a system of algebraic equations:

$$\sum_{n=1}^{N_w}\sum_{i=1}^{N_g}[Z]_{mn}^{ji}\{I\}_{ni} = 0, \quad m = 1,2,...,N_w \tag{15}$$

where $N_w$ is the total number of wires, $N_g$ is total number of elements along the wire grid. Mutual impedance $[Z]_{mn}^{ji}$ represents the interaction between different wire segments is given by:

$$[Z]_{mn}^{ji} = -\frac{1}{j4\pi\omega\varepsilon_{eff}}\begin{bmatrix} \int_{-1}^{1}\int_{-1}^{1}\{D\}_j\{D'\}_i^T g_{0,mn}(s_m,s_n')ds_n'ds_m + \\ +k_1^2\int_{-1}^{1}\int_{-1}^{1}\vec{s}_m\cdot\vec{s}_n'\cdot\{f\}_j\{f'\}_i^T g_{0,mn}(s_m,s_n')ds_n'ds_m \\ -\Gamma_{ref}\int_{-1}^{1}\int_{-1}^{1}\{D\}_j\{D'\}_i^T g_{i,mn}(s_m,s_n^*)ds_n'ds_m^* + \\ +\Gamma_{ref}k_1^2\int_{-1}^{1}\int_{-1}^{1}\vec{s}_m\cdot\vec{s}_n^*\cdot\{f\}_j\{f'\}_i^T g_{i,mn}(s_m,s_n^*)ds_n'ds_m^* \end{bmatrix} \tag{16}$$

Matrices $\{f\}$ and $\{f'\}$ contain the shape functions while $\{D\}$ and $\{D'\}$ contain their derivatives, and $\Delta l_i$, $\Delta l_j$ are the widths of $i$-th and $j$-th boundary elements. The implementation of isoparametric elements gives:

$$[Z]_{mn}^{ji} = -\frac{1}{j4\pi\omega\varepsilon_{eff}}\begin{bmatrix} \int_{-1}^{1}\int_{-1}^{1}\{D\}_j\{D'\}_i^T g_{0,mn}(s_m,s_n')\frac{ds_n'}{d\zeta'}d\zeta'\frac{ds_m}{d\zeta}d\zeta + \\ +k_1^2\int_{-1}^{1}\int_{-1}^{1}\vec{s}_m\cdot\vec{s}_n'\cdot\{f\}_j\{f'\}_i^T g_{0,mn}(s_m,s_n')\frac{ds_n'}{d\zeta'}d\zeta'\frac{ds_m}{d\zeta}d\zeta \\ -\Gamma_{ref}\int_{-1}^{1}\int_{-1}^{1}\{D\}_j\{D'\}_i^T g_{i,mn}(s_m,s_n^*)\frac{ds_n'}{d\zeta'}d\zeta'\frac{ds_m}{d\zeta}d\zeta + \\ +\Gamma_{ref}k_1^2\int_{-1}^{1}\int_{-1}^{1}\vec{s}_m\cdot\vec{s}_n^*\cdot\{f\}_j\{f'\}_i^T g_{i,mn}(s_m,s_n^*)\frac{ds_n'}{d\zeta'}d\zeta'\frac{ds_m}{d\zeta}d\zeta \end{bmatrix} \tag{17}$$

The current source $I_g$ is taken into account as a forced boundary condition at the certain node $i$ of the grounding system:

$$I_i = I_g \tag{18}$$

The treatment of wire junctions is related to the Kirchhoff's current law in its integral and differential form, respectively, related to continuity of currents and charges at the junction.

## 4 Numerical results

Some typical grounding configurations regarding two-wire junctions and three-wire junctions are analyzed. In all cases the unit current source is given by:

$$I_g = 1e^{j0} \tag{19}$$

### 4.1 Two-wire junction

The grounding system in the form of two wire junction is shown in Fig. 2. In Figs 3 to 7 the current distribution along the two horizontal grounding electrodes is presented.

Fig. 3 is related to two 6m long electrodes ($L_1$=$L_2$=6m), with radius $a$= 5mm, the burial depth $d$=0.5m, ground permittivity $\varepsilon_r$=10, and specific resistance ρ= of 100Ωm. The frequency is $f$=10MHz.

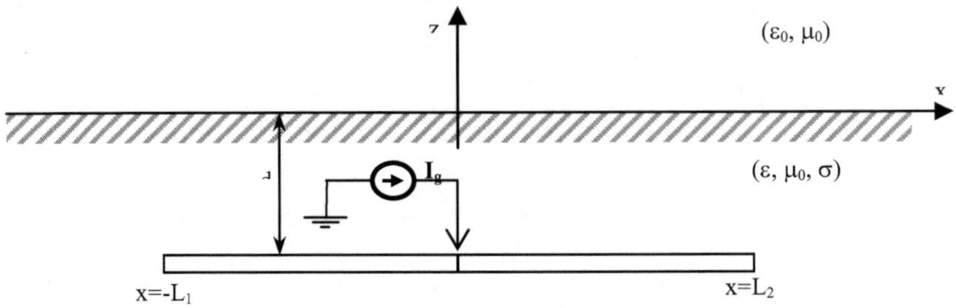


Figure 2: Two-wire junction energized by the current source.

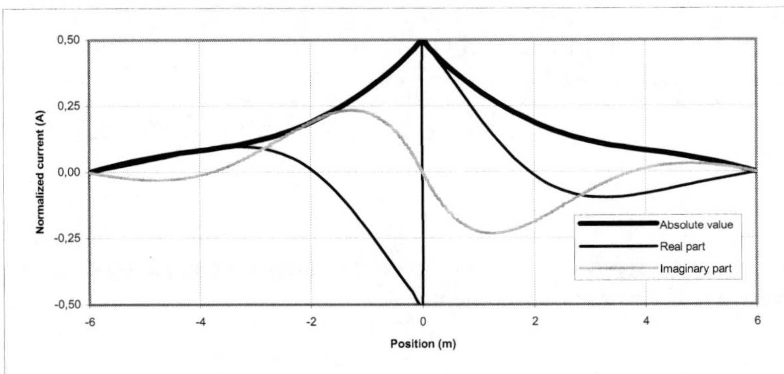


Figure 3: The current distribution along the two-wire junction: $L_1$=6m, $L_2$=6m, $d$=0.5m, $\varepsilon_r$=10, ρ=100Ωm, $f$=10MHz.

Fig. 4 shows the current distribution along the two electrodes of different length ($L_1$=4m, $L_2$=8m)while the other parameters are the same.

Figure 5 shows the induced current distribution along the two 6m long electrodes for the case of higher specific resistance ρ=5400Ωm, while the other parameters are the same.

Figures 6 and 7 show the induced current distribution along the two electrodes of different length; $L_1$=4m, $L_2$=8m, and $L_1$=3m, $L_2$=9m, the specific resistance is ρ=5400Ωm, while the other parameters are the same.

The effect of the different electrode lengths and different values of ground specific resistance is clearly visible.

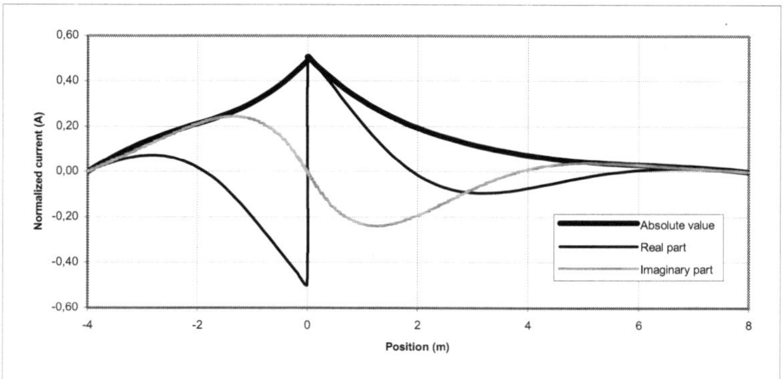


Figure 4: The current distribution along the two-wire junction: $L_1$=4m, $L_2$=8m, $d$=0.5m, $\varepsilon_r$=10, ρ=100Ωm, $f$=10MHz

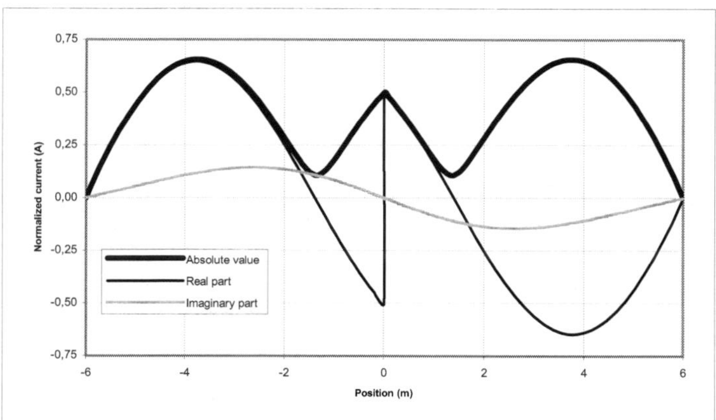


Figure 5: The current distribution along the two-wire junction: $L_1$=6m, $L_2$=6m, $d$=0.5m, $\varepsilon_r$=10, ρ=5400Ωm, $f$=10MHz.

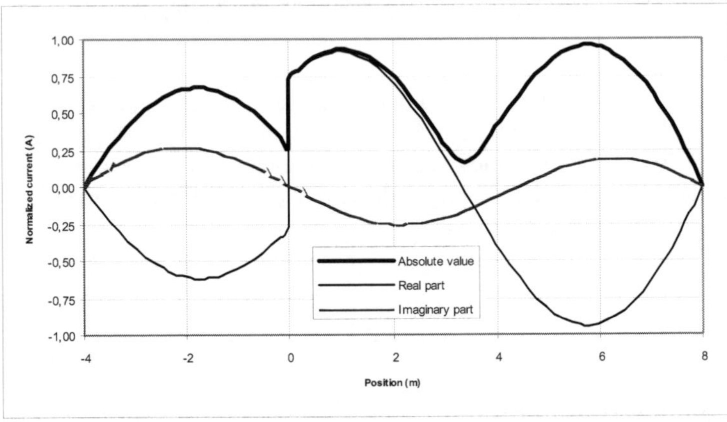


Figure 6: The current distribution along the two-wire junction: $L_1$=4m, $L_2$=8m, $d$=0.5m, $\varepsilon_r$=10, $\rho$=5400Ωm, $f$=10MHz.

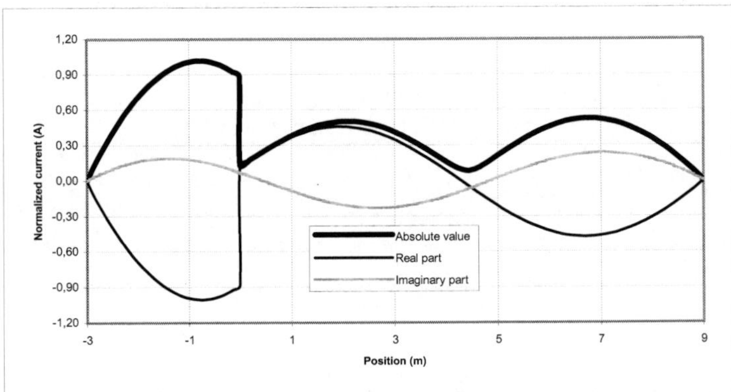


Figure 7: The current distribution along the two-wire junction: $L_1$=3m, $L_2$=9m, $d$=0.5m, $\varepsilon_r$=10, $\rho$=5400Ωm, $f$=10MHz.

### 4.2 The three-wire (T and Y) junctions

Next set of computational examples is related to grounding systems in the form of T-junction and Y-junction, Fig. 8. The ground permittivity is $\varepsilon_r$=10, specific resistance $\rho$= 1000Ωm, the burial depth $d$=0.5m and the frequency is $f$=10MHz.

The current distribution along the T-shaped grounding is shown in Fig. 9, while the Y-shaped grounding system is presented in Fig. 10.

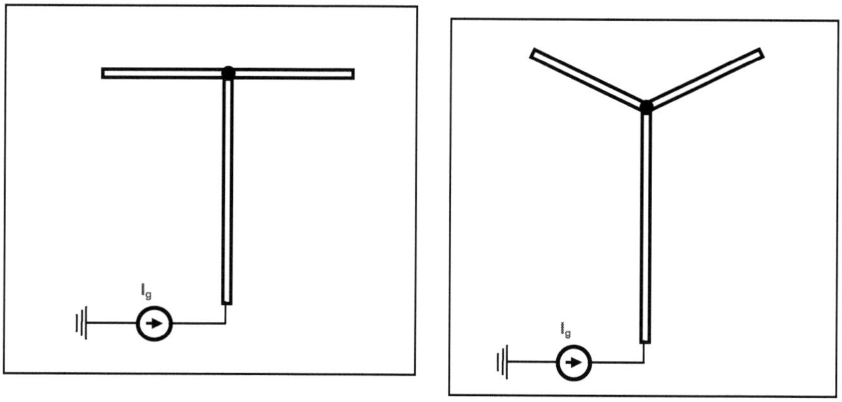


a) T- junction b) Y-junction

Figure 8: The three wire junction energized by th ecurrent source.

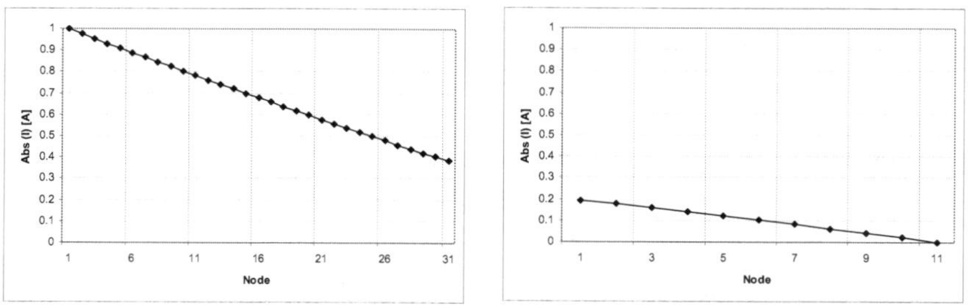


a) $L_1$=1.5m b) $L_2$=0.5m

Figure 9: The current distribution along the T- junction.

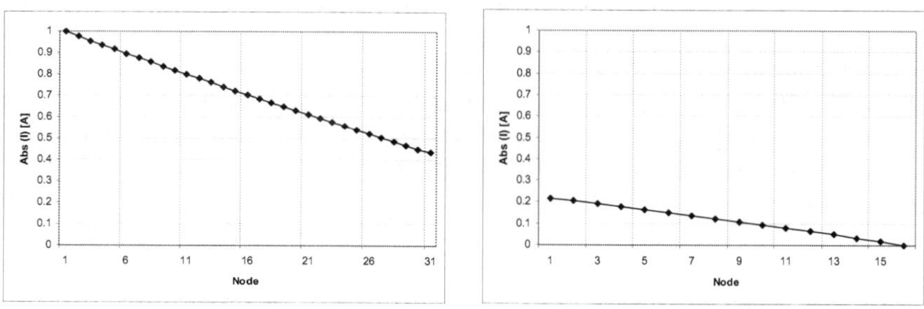


a) $L_1$=1.5m b) $L_2$=0.56m

Figure 10: The current distribution along the Y- junction.

## 5 Concluding remarks

The paper deals with an assessment of the current distribution along the complex grounding systems composed from interconnected conductors. The formulation is based on the wire antenna theory and the related set of Pocklington integro-differential equations for wires of arbitrary shape. The set of Pocklington equations are numerically handled by using the Galerkin-Bubnov scheme of the Indirect Boundary Element Method (GB-IBEM). Some illustrative numerical results are presented in the paper.

Future work will involve the calculation of scattered voltage, input impedance and transient impedance for the case of complex grounding systems including interconnected conductors.

## References

[1] Y. Liu, M. Zitnik, R. Thottappillil, "An Improved Transmission Line Model of Grounding System", IEEE Trans. EMC, Vol.43, No.3, pp. 348-355, 2001.

[2] G. Ala, M. L. Di Silvestre, "A Simulation Model for Electromagnetic Transients in Lightning Protection Systems", IEEE Trans. EMC, Vol.44, No.4, pp.539-534, 2003.

[3] L. Grcev , F. Dawalibi, "An Electromagnetic Model for Transients in Grounding Systems", IEEE Trans. Power Delivery, No 4., pp. 1773-1781, Oct. 1990.

[4] L. D. Grcev, F.E. Menter, "Transient Electro-magnetic Fields Near Large Earthing Systems", IEEE Trans. Magnetics, Vol. 32, pp. 1525-1528, May 1996.

[5] D. Poljak, V. Roje, "The Integral equation method for ground wire impedance", Constanda, C., Saranen, J., Seikkala, S. (Ed), Integral methods in science and engineering, Vol. I, Longman, UK., 139-143, 1997.

[6] D. Poljak, El Khamlichi Drissi, R.Goic, Boundary element modelling of horizontal grounding electrodes using the set of generalized telegrapher's equations, BEM 31, Southampton, UK, Sept. 2009.

[7] D.Poljak, V. Dorić, K. El Khamlichi Drissi, K Kerroum, I. Medić, Comparison of wire antenna and modified transmission line approach to the assessment of frequency response of horizontal grounding electrodes. // *Engineering Analysis with Boundary Elements*. 32 (2008) , 8; 676-681.

[8] R.G. Olsen, M.C. Willis, "A Comparison of Exact and Quasi-static Methods for Evaluating Grounding Systems at High Frequencies", IEEE Trans. Power Delivery, Vol. 11, No 2, pp. 1071-1081, April 1996.

[9] D.Poljak, "Advanced Modeling in Computational EMC ", Wiley, New York, 2007.

# Rotational symmetry applied to boundary element computation for nuclear fusion plasma

M. Itagaki[1], T. Ishimaru[1] & K. Watanabe[2]
[1]*Faculty of Engineering, Hokkaido University, Japan*
[2]*National Institute for Fusion Science, Japan*

## Abstract

An efficient boundary element technique has been proposed to deal with a problem when geometry and boundary conditions have a rotational symmetry. In this technique, no boundary elements need to be defined on the symmetry surface. This idea has been applied to a problem of magnetic field in nuclear fusion plasma. A useful "linear transformation" for the rotational symmetry in the toroidal direction has been derived in terms of the x-, y- and z-components of vector potential in a 3-D Cartesian coordinate system. Applying this linear transformation to the discretized set of boundary integral equations, one can drastically reduce the size of matrix, the number of unknowns and then computing cost. Also, accurate solutions can be expected. Results of test calculations demonstrate the validity of the present formulation.
*Keywords: boundary element method, rotational symmetry, vector potential, linear transformation, nuclear fusion plasma, Cauchy condition surface method, magnetic sensor.*

## 1 Introduction

The present work is a part of the authors' research plan to develop an inverse analytic technique to identify the boundary shape of nuclear fusion plasma from signals of magnetic sensors located outside the plasma. For this purpose, the Cauchy condition surface (CCS) method [1] has already been established for a tokamak-type fusion device. The geometry of tokamak plasma is axisymmetric so that the analysis can be made in a 2-dimensional (2-D), *r-z* system. On the other hand, 3-D analyses are required for non-axisymmetric plasma, e.g., in a helical type device such as the Large Helical Device (LHD). The application of

WIT Transactions on Modelling and Simulation, Vol 50, © 2010 WIT Press
www.witpress.com, ISSN 1743-355X (on-line)
doi:10.2495/BE1000121

the CCS method to a 3-D space analysis, however, becomes quite challenging because the 3-D system has a much larger number of unknowns than a 2-D system.

The authors have noticed that the LHD has a 1/5 rotational symmetry. In Section 2 of the present paper, an efficient boundary element technique is given for solving a simple potential problem having rotational symmetry. When there is an *n*-fold rotational symmetry, the system matrix becomes a cyclic one. Because of this, the number of unknowns is reduced to 1/*n*. In this case, no boundary elements need to be defined on the symmetry surface.

In Section 3, this idea is extended to nuclear fusion plasma that has a rotational symmetry in the toroidal direction. A useful "linear transformation" is derived in terms of the x-, y- and z-components of vector potential in a Cartesian coordinate system. Applying this linear transformation to the discretized set of boundary integral equations, one can reduce the number of unknowns and then computing cost dramatically. Also, accurate solutions can be expected.

Numerical examples are shown in Section 4 to demonstrate the validity of the present formulation to deal with the rotational symmetry.

## 2 Rotational symmetry in a simple potential problem

First of all, one here considers a simple potential problem. Corresponding to the 2-D Laplace equation

$$\frac{\partial^2 u}{\partial x^2}+\frac{\partial^2 u}{\partial y^2}=0, \tag{1}$$

the boundary integral equation is given by

$$c_i u_i = \int_\Gamma \left( u_i^* \frac{\partial u}{\partial n} - u \frac{\partial u_i^*}{\partial n} \right) d\Gamma, \tag{2}$$

which discretized form can be written in a matrix equation form [2]

$$\mathbf{Hu} = \mathbf{Gq}. \tag{3}$$

Here, the quantities $\mathbf{u}$ and $\mathbf{q}$ denote vectors of the Dirichlet condition $u$ and the Neumann condition $q = \partial u / \partial n$, respectively on the boundary $\Gamma$.

When the boundary $\Gamma$ is divided into $n$ segments, ($\Gamma_1, \Gamma_2, \cdots, \Gamma_n$), eqn (3) is given by

$$\begin{bmatrix} \mathbf{H}_{1,1} & \mathbf{H}_{1,2} & \cdots & \mathbf{H}_{1,n} \\ \mathbf{H}_{2,1} & \mathbf{H}_{2,2} & \cdots & \mathbf{H}_{2,n} \\ \cdots & \cdots & \cdots & \cdots \\ \mathbf{H}_{n,1} & \mathbf{H}_{n,2} & \cdots & \mathbf{H}_{n,n} \end{bmatrix} \begin{Bmatrix} \mathbf{u}_1 \\ \mathbf{u}_2 \\ \cdots \\ \mathbf{u}_n \end{Bmatrix} = \begin{bmatrix} \mathbf{G}_{1,1} & \mathbf{G}_{1,2} & \cdots & \mathbf{G}_{1,n} \\ \mathbf{G}_{2,1} & \mathbf{G}_{2,2} & \cdots & \mathbf{G}_{2,n} \\ \cdots & \cdots & \cdots & \cdots \\ \mathbf{G}_{n,1} & \mathbf{G}_{n,2} & \cdots & \mathbf{G}_{n,n} \end{bmatrix} \begin{Bmatrix} \mathbf{q}_1 \\ \mathbf{q}_2 \\ \cdots \\ \mathbf{q}_n \end{Bmatrix} \tag{4}$$

using submatrices $\mathbf{H}_{i,j}$ and $\mathbf{G}_{i,j}$. If there is an *n*-fold rotational symmetry in the geometry as well as the boundary conditions under consideration, one finds

$\mathbf{u}_1 = \mathbf{u}_2 = \cdots = \mathbf{u}_n$ and $\mathbf{q}_1 = \mathbf{q}_2 = \cdots = \mathbf{q}_n$. Also, the matrix on each side in eqn (4) becomes a cyclic one. That is, eqn (4) can be rewritten as

$$\begin{bmatrix} \mathbf{H}_{1,1} & \mathbf{H}_{1,2} & \cdots & \mathbf{H}_{1,n} \\ \mathbf{H}_{1,n} & \mathbf{H}_{1,1} & \cdots & \mathbf{H}_{1,n-1} \\ \cdots & \cdots & \cdots & \cdots \\ \mathbf{H}_{1,2} & \mathbf{H}_{1,3} & \cdots & \mathbf{H}_{1,1} \end{bmatrix} \begin{Bmatrix} \mathbf{u}_1 \\ \mathbf{u}_1 \\ \cdots \\ \mathbf{u}_1 \end{Bmatrix} = \begin{bmatrix} \mathbf{G}_{1,1} & \mathbf{G}_{1,2} & \cdots & \mathbf{G}_{1,n} \\ \mathbf{G}_{1,n} & \mathbf{G}_{1,1} & \cdots & \mathbf{G}_{1,n-1} \\ \cdots & \cdots & \cdots & \cdots \\ \mathbf{G}_{1,2} & \mathbf{G}_{1,3} & \cdots & \mathbf{G}_{1,1} \end{bmatrix} \begin{Bmatrix} \mathbf{q}_1 \\ \mathbf{q}_1 \\ \cdots \\ \mathbf{q}_1 \end{Bmatrix}. \tag{5}$$

Equation (5) can then be simplified as

$$\left[\mathbf{H}_{1,1} + \mathbf{H}_{1,2} + \cdots + \mathbf{H}_{1,n}\right]\{\mathbf{u}_1\} = \left[\mathbf{G}_{1,1} + \mathbf{G}_{1,2} + \cdots + \mathbf{G}_{1,n}\right]\{\mathbf{q}_1\}. \tag{6}$$

Now the matrix size and the number of unknowns are reduced to $1/n^2$ and $1/n$, respectively. It should be noted that in this case one need not to define any boundary elements on the symmetry surface, so that the periodic boundary conditions are naturally satisfied.

Suppose now $J$ mesh points, $(x_j^{(1)}, y_j^{(1)}),\ (j = 1, 2, \cdots, J)$, have been given for the boundary $\Gamma_1$ to generate the submatrices $\mathbf{H}_{1,1}$ and $\mathbf{G}_{1,1}$. Other mesh point coordinates

$$(x_j^{(k)}, y_j^{(k)}),\ (k = 2, 3, \cdots, n;\ j = 1, 2, \cdots, J)$$

to compute $[\mathbf{H}_{1,2}, \mathbf{H}_{1,3}, \cdots, \mathbf{H}_{1,n}]$ and $[\mathbf{G}_{1,2}, \mathbf{G}_{1,3}, \cdots, \mathbf{G}_{1,n}]$ are calculated using the well-known linear transformation

$$\begin{Bmatrix} x_j^{(k)} - x_0 \\ y_j^{(k)} - y_0 \end{Bmatrix} = \begin{bmatrix} \cos(k-1)\alpha & -\sin(k-1)\alpha \\ \sin(k-1)\alpha & \cos(k-1)\alpha \end{bmatrix} \begin{Bmatrix} x_j^{(1)} - x_0 \\ y_j^{(1)} - y_0 \end{Bmatrix} \tag{7}$$

with the angle $\alpha = 2\pi/n$ and the rotation axis $(x_0, y_0)$, as shown in fig. 1.

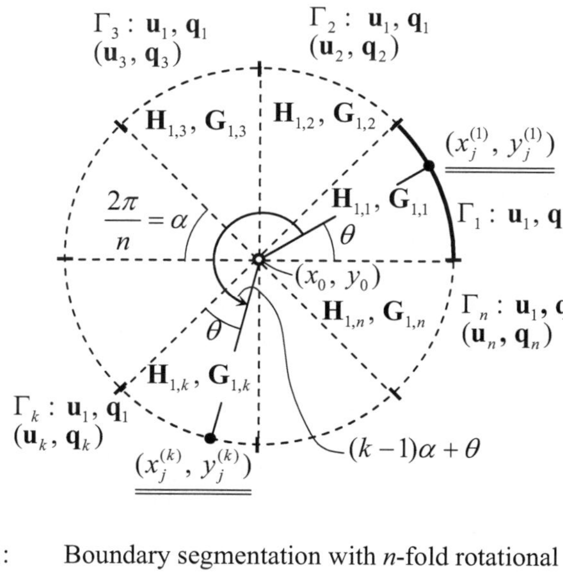


Figure 1: Boundary segmentation with $n$-fold rotational symmetry.

# 3 Rotational symmetry of 3-D vector potential

The fundamental idea described in Section 2 is now applied to a problem of magnetic field in nuclear fusion plasma, which is described using vector potential in a 3-D space. This section describes how the rotational symmetry is introduced to reduce the number of unknowns in the 3-D CCS method formulation that is given by a set of boundary integral equations for points along the CCS and for the magnetic sensor positions. The boundary integral equations formulated in this work are all described in a Cartesian coordinate system. The reason for this is given in the Appendix. An outline of 3-D CCS method is found in the literature [3].

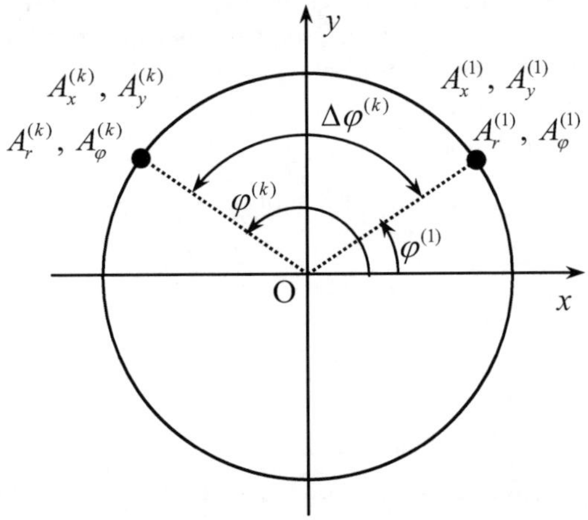


Figure 2: Rotational transformation of vector potentials.

## 3.1 Linear transformation of vector potential

Now one describes the vector potential $(A_x^{(k)}, A_y^{(k)}, A_z^{(k)})$ in the $k$-th segment in terms of the vector potential $(A_x^{(1)}, A_y^{(1)}, A_z^{(1)})$ in the first segment. The relationship between $(A_x^{(k)}, A_y^{(k)})$ and $(A_r^{(k)}, A_\varphi^{(k)})$ in the different coordinate systems are described using the toroidal angle $\varphi^{(k)}$, as

$$\begin{pmatrix} A_x^{(k)} \\ A_y^{(k)} \end{pmatrix} = \begin{pmatrix} \cos\varphi^{(k)} & -\sin\varphi^{(k)} \\ \sin\varphi^{(k)} & \cos\varphi^{(k)} \end{pmatrix} \begin{pmatrix} A_r^{(k)} \\ A_\varphi^{(k)} \end{pmatrix}. \tag{8}$$

Considering the rotational symmetry, i.e., $A_r^{(k)} = A_r^{(1)}$ and $A_\varphi^{(k)} = A_\varphi^{(1)}$, one finds

$$\begin{pmatrix} A_x^{(k)} \\ A_y^{(k)} \end{pmatrix} = \begin{pmatrix} \cos\varphi^{(k)} & -\sin\varphi^{(k)} \\ \sin\varphi^{(k)} & \cos\varphi^{(k)} \end{pmatrix} \begin{pmatrix} A_r^{(1)} \\ A_\varphi^{(1)} \end{pmatrix}. \tag{9}$$

With the relationship between $(A_r^{(1)},\ A_\varphi^{(1)})$ and $(A_x^{(1)},\ A_y^{(1)})$, eqn (9) becomes

$$\begin{pmatrix} A_x^{(k)} \\ A_y^{(k)} \end{pmatrix} = \begin{pmatrix} \cos\varphi^{(k)} & -\sin\varphi^{(k)} \\ \sin\varphi^{(k)} & \cos\varphi^{(k)} \end{pmatrix} \begin{pmatrix} \cos\varphi^{(1)} & \sin\varphi^{(1)} \\ -\sin\varphi^{(1)} & \cos\varphi^{(1)} \end{pmatrix} \begin{pmatrix} A_x^{(1)} \\ A_y^{(1)} \end{pmatrix}. \tag{10}$$

This formula can be transformed into

$$\begin{pmatrix} A_x^{(k)} \\ A_y^{(k)} \end{pmatrix} = \begin{pmatrix} \cos\Delta\varphi^{(k)} & -\sin\Delta\varphi^{(k)} \\ \sin\Delta\varphi^{(k)} & \cos\Delta\varphi^{(k)} \end{pmatrix} \begin{pmatrix} A_x^{(1)} \\ A_y^{(1)} \end{pmatrix} \tag{11}$$

using the "additional theorem" of trigonometric functions with a rotation angle

$$\Delta\varphi^{(k)} = \varphi^{(k)} - \varphi^{(1)}.$$

Including the $z$-component, one finally obtains

$$\begin{pmatrix} A_x^{(k)} \\ A_y^{(k)} \\ A_z^{(k)} \end{pmatrix} = \begin{pmatrix} \cos\Delta\varphi^{(k)} & -\sin\Delta\varphi^{(k)} & 0 \\ \sin\Delta\varphi^{(k)} & \cos\Delta\varphi^{(k)} & 0 \\ 0 & 0 & 1 \end{pmatrix} \begin{pmatrix} A_x^{(1)} \\ A_y^{(1)} \\ A_z^{(1)} \end{pmatrix}. \tag{12}$$

This is the "linear transformation" from $(A_x^{(1)},\ A_y^{(1)},\ A_z^{(1)})$ to $(A_x^{(k)},\ A_y^{(k)},\ A_z^{(k)})$ in a Cartesian coordinate system.

### 3.2 Application to boundary element equations in 3-D CCS method

The set of boundary integral equations for the 3-D CCS method can also be transformed into a matrix equation

$$\mathbf{Hu} = \mathbf{Gq}, \tag{13}$$

where

$$\mathbf{u} = \left\{ \mathbf{A}_x^{(1)}, \mathbf{A}_y^{(1)}, \mathbf{A}_z^{(1)}, \cdots, \mathbf{A}_x^{(n)}, \mathbf{A}_y^{(n)}, \mathbf{A}_z^{(n)} \right\}^T \tag{14}$$

and

$$\mathbf{q} = \left\{ \frac{\partial \mathbf{A}_x^{(1)}}{\partial n}, \frac{\partial \mathbf{A}_y^{(1)}}{\partial n}, \frac{\partial \mathbf{A}_z^{(1)}}{\partial n}, \cdots, \frac{\partial \mathbf{A}_x^{(n)}}{\partial n}, \frac{\partial \mathbf{A}_y^{(n)}}{\partial n}, \frac{\partial \mathbf{A}_z^{(n)}}{\partial n} \right\}^T. \tag{15}$$

Here, $(\mathbf{A}_x^{(k)},\ \mathbf{A}_y^{(k)},\ \mathbf{A}_z^{(k)})$ means the set of all vector potentials within the $k$-th segment. The LHS of eqn (13), for example, can be written as

$$\mathbf{Hu} = \begin{bmatrix} \mathbf{H}_{11} & \mathbf{H}_{12} & \mathbf{H}_{13} & \cdots & \mathbf{H}_{1,3n-2} & \mathbf{H}_{1,3n-1} & \mathbf{H}_{1,3n} \\ \mathbf{H}_{21} & \mathbf{H}_{22} & \mathbf{H}_{23} & \cdots & \mathbf{H}_{2,3n-2} & \mathbf{H}_{2,3n-1} & \mathbf{H}_{2,3n} \\ \mathbf{H}_{31} & \mathbf{H}_{32} & \mathbf{H}_{33} & \cdots & \mathbf{H}_{3,3n-2} & \mathbf{H}_{3,3n-1} & \mathbf{H}_{3,3n} \\ \vdots & \vdots & \vdots & & \vdots & \vdots & \vdots \\ \mathbf{H}_{l,1} & \mathbf{H}_{l,2} & \mathbf{H}_{l,3} & & \mathbf{H}_{l,3n-2} & \mathbf{H}_{l,3n-1} & \mathbf{H}_{l,3n} \\ \vdots & \vdots & \vdots & & \vdots & \vdots & \vdots \\ \mathbf{H}_{L-1,1} & \mathbf{H}_{L-1,2} & \mathbf{H}_{L-1,3} & \cdots & \mathbf{H}_{L-1,3n-2} & \mathbf{H}_{L-1,3n-1} & \mathbf{H}_{L-1,3n} \\ \mathbf{H}_{L,1} & \mathbf{H}_{L,2} & \mathbf{H}_{L,3} & \cdots & \mathbf{H}_{L,3n-2} & \mathbf{H}_{L,3n-1} & \mathbf{H}_{L,3n} \end{bmatrix} \begin{pmatrix} \mathbf{A}_x^{(1)} \\ \mathbf{A}_y^{(1)} \\ \mathbf{A}_z^{(1)} \\ \vdots \\ \mathbf{A}_x^{(n)} \\ \mathbf{A}_y^{(n)} \\ \mathbf{A}_z^{(n)} \end{pmatrix}, \tag{16}$$

 
www.witpress.com, ISSN 1743-355X (on-line)

where $L$ denotes the number of lines that depends on the number of nodal points on the CCS and the number of magnetic sensors as well. Assuming an $n$-fold rotational symmetry, i.e., applying eqn (12) to the vector on the RHS in eqn (16), eqn (16) is described only using $\mathbf{A}_x^{(1)}$, $\mathbf{A}_y^{(1)}$ and $\mathbf{A}_z^{(1)}$. For instance, the $l$ th line of $\mathbf{Hu}$ can be written in the form

$$(\mathbf{Hu})_l = \begin{bmatrix} \sum_{k=1}^{n}\left(\mathbf{H}_{l,3k-2}\cos\Delta\varphi^{(k)} + \mathbf{H}_{l,3k-1}\sin\Delta\varphi^{(k)}\right), \\ \sum_{k=1}^{n}\left(-\mathbf{H}_{l,3k-2}\sin\Delta\varphi^{(k)} + \mathbf{H}_{l,3k-1}\cos\Delta\varphi^{(k)}\right), \\ \sum_{k=1}^{n}\mathbf{H}_{l,3k} \end{bmatrix}^{T} \begin{pmatrix} \mathbf{A}_x^{(1)} \\ \mathbf{A}_y^{(1)} \\ \mathbf{A}_z^{(1)} \end{pmatrix}. \quad (17)$$

The $\mathbf{Gq}$ on the RHS of eqn (13) can also be rewritten in the same way. Both the numbers of columns and unknowns can be reduced to $1/n$ if only giving the rotational angle $\Delta\varphi^{(k)}$.

When a singular point $i$ is located on the CCS, the boundary integral equation for each of $\mathbf{A}_x$, $\mathbf{A}_y$ and $\mathbf{A}_z$ holds independently, and all equations commonly use the same fundamental solution. Because of this, in eqn (16) the portion related to the CCS represents a $3n \times 3n$ square matrix equation:

$$\mathbf{Hu}^{CCS} = \begin{bmatrix} \mathbf{H}_{11} & 0 & 0 & \cdots & \mathbf{H}_{1,3n-2} & 0 & 0 \\ 0 & \mathbf{H}_{22} & 0 & \cdots & 0 & \mathbf{H}_{2,3n-1} & 0 \\ 0 & 0 & \mathbf{H}_{33} & \cdots & 0 & 0 & \mathbf{H}_{3,3n} \\ \vdots & \vdots & \vdots & & \vdots & \vdots & \vdots \\ \mathbf{H}_{3n-2,1} & 0 & 0 & \cdots & \mathbf{H}_{3n-2,3n-2} & 0 & 0 \\ 0 & \mathbf{H}_{3n-1,2} & 0 & \cdots & 0 & \mathbf{H}_{3n-1,3n-1} & 0 \\ 0 & 0 & \mathbf{H}_{3n,3} & \cdots & 0 & 0 & \mathbf{H}_{3n,3n} \end{bmatrix} \begin{pmatrix} \mathbf{A}_x^{(1)} \\ \mathbf{A}_y^{(1)} \\ \mathbf{A}_z^{(1)} \\ \vdots \\ \mathbf{A}_x^{(n)} \\ \mathbf{A}_y^{(n)} \\ \mathbf{A}_z^{(n)} \end{pmatrix}. \quad (18)$$

Then, introducing the rotational symmetry, eqn (18) can be simplified as

$$\mathbf{Hu}_{123}^{CCS} = \begin{bmatrix} \sum_{k=1}^{n}\left(\mathbf{H}_{1,3k-2}\cos\Delta\varphi^{(k)}\right) & \sum_{k=1}^{n}\left(-\mathbf{H}_{1,3k-2}\sin\Delta\varphi^{(k)}\right) & 0 \\ \sum_{k=1}^{n}\left(\mathbf{H}_{2,3k-1}\sin\Delta\varphi^{(k)}\right) & \sum_{k=1}^{n}\left(\mathbf{H}_{2,3k-1}\cos\Delta\varphi^{(k)}\right) & 0 \\ 0 & 0 & \sum_{k=1}^{n}\mathbf{H}_{3,3k} \end{bmatrix} \begin{pmatrix} \mathbf{A}_x^{(1)} \\ \mathbf{A}_y^{(1)} \\ \mathbf{A}_z^{(1)} \end{pmatrix}. \quad (19)$$

Finally, the equations for points on the CCS as well as the ones for the magnetic sensor positions are solved simultaneously. Once all the values on the CCS are known, one can calculate the magnetic field for arbitrary points.

# 4 Numerical example

One here considers a problem to model the non-axisymmetric plasma in the LHD, a helical-type device. The three components of magnetic field in the LHD were reconstructed using the 3-D CCS method. As the magnetic field profile is not axisymmetric, this problem is challenging and requires a large number of unknowns.

The results were compared with the reference solution obtained using the HINT code [4]. Figure 3 shows an example of the reference solution obtained using the HINT code. This figure gives the contours of the $r$-component of the magnetic field $B_r$ on the $r$-$z$ plane at the toroidal angle of 18 deg.

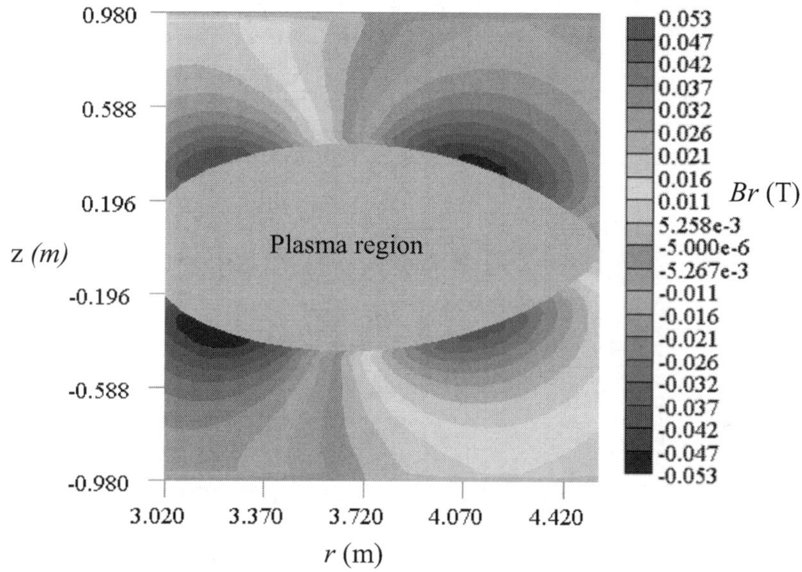


Figure 3: Reference profile of magnetic field $B_r$ at $\varphi = 18$ deg.

For inverse analyses, one here assumes that 20 magnetic flux loop sensors and 451 magnetic field sensors are arranged outside the plasma. Each of field sensors is hypothetically assumed to detect all of the three components of magnetic field. In this case the number of sensor signals is $451 \times 3 + 20 = 1373$. A tube-shaped CCS was placed within a domain that can be supposed to be inside the actual plasma.

First, a CCS that covers 360-deg. was divided into 160 boundary elements (Case A). As the number of unknowns, 8640, is larger than the number of equations, 5693, as shown in Table 1, it was impossible to solve this problem.

Next, considering a 1/5 rotational symmetry, only 72-deg. portion of the CCS tube was modelled and this portion was divided into 32 boundary elements (Case B). Figure 4 shows the reconstructed contours of the *r*-component of the magnetic field $B_r$ on the *r-z* plane at the toroidal angle of 18 deg., which correspond to the reference contours in fig. 3. As the magnetic fields computed using the CCS method has no physical meaning inside the plasma boundary [3], they are not drawn inside the plasma in fig. 4. This reconstructed field profile agrees well with the reference one in fig. 3.

Table 1: Calculation condition for plasma in LHD.

| Case | A | B | C |
|---|---|---|---|
| Rotational symmetry | 360-deg. | 1/5 symmetry | 360-deg. |
| No. of sensor signals | 1373 | 1373 | 1373 |
| No. of boundary elements | 160 | 32 | 40 |
| No. of equations | 5693 | 2237 | 2453 |
| No. of unknowns | 8640 | 1728 | 2160 |
| Solution accuracy | Not solved | Acceptable | Poor |

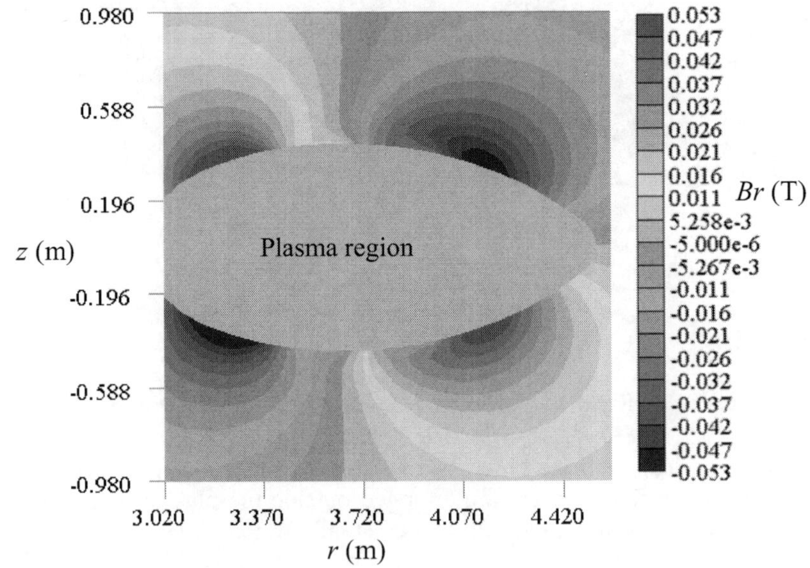


Figure 4: Profile of reconstructed magnetic field $B_r$ at $\varphi = 18$ deg. (Case B).

When one took only 40 boundary elements on the CCS for 360-deg. (Case C), the number of unknowns is of course smaller than that of equations. Unfortunately, however the accuracy of the reconstructed solution in this case was very poor, as shown in fig. 5. That is, without considering the rotational symmetry, it is difficult to realize a good accuracy in the 3-D analysis of non-axisymmetric plasma, which requires a large number of unknowns.

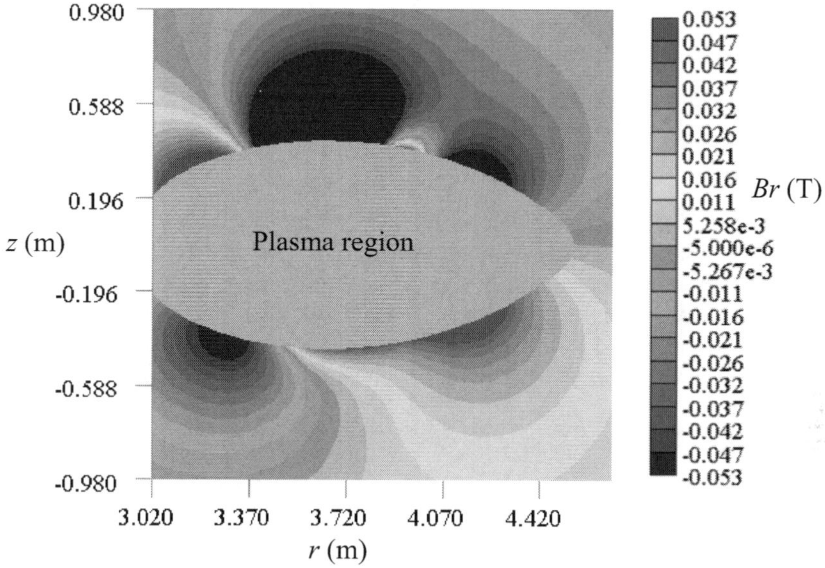


Figure 5: Profile of reconstructed magnetic field $B_r$ at $\varphi = 18$ deg. (Case C).

# 5 Conclusion

An efficient boundary element technique has been applied to problems of magnetic field in 3-D nuclear fusion plasma that has a rotational symmetry in the toroidal direction. In this technique, no boundary elements need to be defined on the symmetry surface. The linear transformation in terms of the x-, y- and z-components of 3-D vector potential plays an important role to reduce the matrix size and the number of unknowns in the discretized set of boundary integral equations. This enables one to realize not only the reduction of the computing cost but the improvement of numerical solution accuracy.

Results of test calculations for nuclear fusion plasmas in a helical-type device demonstrate the validity of the present formulation. It has been found that one should incorporate the rotational symmetry into the analysis of non-axisymmetric 3-D plasma in order to obtain an acceptable accuracy of the solution.

## Acknowledgements

This research was supported by the Ministry of Education, Culture, Sports, Science and Technology, Grant-in-Aid for Scientific Research (C), 20560762, 2009. This work was also performed with the support and under the auspices of the NIFS Collaboration Research Program (NIFS08KLHH308).

## Appendix: Why the boundary integral equations in this work are described in a Cartesian coordinate system?

The vector Laplacian in a Cartesian coordinate system has a simple relationship

$$(\nabla^2 \boldsymbol{A})_k = \nabla^2 A_k \quad (k = x,\, y,\, z)\,. \tag{A1}$$

That is, the vector Laplacian can be given by a set of the scalar Laplacian of each Cartesian scalar component. In a cylindrical or a spherical coordinate system, on the other hand, the expression of the vector Laplacian is not so straightforward. In a cylindrical system, for example, the components of the vector Laplacian are written in complicated forms, as

$$(\nabla^2 \boldsymbol{A})_r = \nabla^2 A_r - \frac{2}{r^2}\frac{\partial A_\varphi}{\partial \varphi} - \frac{A_r}{r^2}\,, \tag{A2a}$$

$$(\nabla^2 \boldsymbol{A})_\varphi = \nabla^2 A_\varphi + \frac{2}{r^2}\frac{\partial A_r}{\partial \varphi} - \frac{A_\varphi}{r^2} \tag{A2b}$$

and

$$(\nabla^2 \boldsymbol{A})_z = \nabla^2 A_z\ . \tag{A2c}$$

If one uses this coordinate system, the boundary integral equations corresponding to $r$- and $\varphi$- components will include domain integral terms. To realize a boundary-only integral formulation, it is better not to use a cylindrical or a spherical system. Because of this, the authors adopt a 3-D Cartesian coordinate system for the analysis to obtain the 3-D distribution of vector potential. However, it is easy to transform the result, once calculated in a Cartesian coordinate system, into one in another coordinate system.

## References

[1] Kurihara, K, *Fusion Eng. Des.*, **51-52,** pp.1049-1057, 2000.

[2] Brebbia, C.A., "*The Boundary Element Method for Engineers*", Pentech Press, London, 1978

[3] Itagaki, M., Maeda, T., Wakasa, A., Watanabe, K., Proc. BEM/MRM XXXI, Southampton, pp.397-404, 2009.

[4] Harafuji, K., Hayashi T., Sato T., *J. Comput. Phys.,* **81,** pp.169-192, 1989.

# Section 5
# Advanced meshless and mesh reduction methods

# A new generalized meshless method technique for tracheobronchial airflow modeling

K. Erhart[1], E. Divo[1], A. Kassab[1], S. Gerace[1], A. Santhanam[1,2] & P. Kupelian[2]
[1]*University of Central Florida, USA*
[2]*M. D. Anderson Cancer Center, USA*

## Abstract

Numerous approaches to the implementation of Meshless methods have been developed and described over the last several years for a wide range of physical applications. Despite this wide range of applications and implementation approaches, few attempts have been made to combine the various schemes utilized throughout the Meshless research community in the hopes of achieving a more resilient and accurate Meshless method. This paper describes such a generalized Meshless approach that incorporates moving least squares, radial basis functions, and finite differencing into a unique Meshless-based solution framework. The development of this approach was stimulated by stability requirements that arise in the difficult geometries that appear in typical biological flow modeling applications. Therefore, this new approach is described in the context of tracheobronchial flow modeling for the prediction of thoracic tumor movement throughout the breathing cycle.
*Keywords: meshless methods, generalized finite differencing, porous lung flow.*

## 1 Introduction

The medical community is currently giving much attention to the idea of individualized medical treatment. Individualized treatment means that medication, surgery, and other procedures are selected based on the specific details of each patient, rather than through a more general disease based protocol. This process allows the use of drugs and treatments that may be harmful to some patients to be utilized successfully on those patients that have conditions and circumstances ideally suited to the treatment. The area of cancer treatment has been following

WIT Transactions on Modelling and Simulation, Vol 50, © 2010 WIT Press
www.witpress.com, ISSN 1743-355X (on-line)
doi:10.2495/BE100131

this trend for some time, as size and anatomy can vary significantly from patient to patient. In conformal and intensity modulated radiation therapy for example, the exact details of the individual patient anatomy are absolutely critical to successful treatment outcomes. This is especially important for cancers in anatomic regions that experience motion due to breathing, such as the lungs, chest, and abdomen. Therefore much effort has recently gone into the area of real-time (4D) medical imaging. These advanced imaging techniques have made it possible to quantify expected tumor motions throughout the breathing cycle and have provided confidence that radiation is being delivered to the correct cancer site. Despite these improved imaging capabilities, there is currently no process available to visualize tumor motion during the actual course of radiation treatment. Therefore, the medical staff must rely upon previously obtained images to predict the motion during the actual treatment procedure. This research aims to improve these prediction capabilities by correlating real-time airflow data with the lung/tumor motion. A standard spirometer can be used to monitor the patient's breathing rate and computational fluid dynamics and mechanics data are used to predict the resulting lung deformations, thereby providing an improved means of estimating the current tumor position. In addition, the imaging data and computational models can be used to improve the accuracy of lung tissue property estimations so that patient treatment and tissue complications can be more accurately assessed.

This paper will introduce a new generalized Meshless computational modeling approach, describe its application to the area of tracheobronchial flow modeling, and discuss the preliminary findings of the research. The computational modeling has been carried out using a Meshless-based approach for both flow and elasticity analysis. Meshless methods offer an extremely increased ease of use for biological applications as individual patient anatomies are quite complex and therefore very difficult to represent with typical finite element or finite volume meshes. However, as with any automatically discretized technique, solution stability and robustness are extremely key factors to the methodology and therefore have been key aspects in guiding this research.

In order to present this research in the most suitable fashion, this paper is organized into several sections, which are then followed by some overall concluding remarks. In the first section, we will present a more complete description of the research goals, tasks, and approach. This section will describe the entire solution procedure and coupling process as it is not tied specifically to Meshless methods, therefore the ideas and approaches utilized within this work can be understood by those without experience in the field of Meshless method modeling. Following this application overview, a section describing the pertinent details of our new generalized Meshless technique will be presented along with numerous references to more detailed Meshless method publications. Finally, results will be presented to demonstrate the current progress that has been made in individualized patient tracheobronchial airflow modeling for cancer treatment applications.

## 2 Tracheobronchial airflow modeling

Despite large reductions in smoking habits and decreased use of harmful building materials and chemicals, lung cancer is still the second most common type of cancer in the United States. While radiation therapy has made huge strides in offering curative treatments for many types of cancer, its use in treating lung cancer is particularly difficult due to motion of the tumor that is induced by normal patient breathing. Due to the length of the treatment and the relative weakness of many of the lung cancer patients, simply restricting or reducing breathing during treatment is not a viable means of reducing tumor motion (radiation treatment is typically delivered in 5-15 minute doses over multiple consecutive days). Additionally, the motion patterns of many lung tumors can vary unpredictably from day-to-day [1, 2], the depth of breathing can change even in short amounts of time, and there are notable individual differences in breathing and motion patterns from one patient to another. These difficulties complicate treatment and since providing adequate radiation dose to the tumor is the primary goal of radiation therapy, treatment fields are often enlarged to accommodate for these motions and uncertainties. This results in higher doses to normal tissue, thereby increasing the toxicity rates from therapy. Through advanced medical imaging it has been found that lung tumors can move as much as 3 cm while the average tumor size in patients is only $5.5 \pm 3.1$ cm [1], therefore this practice of field enlargement can easily lead to over 50% increases in the field size, much of which would be unnecessary if a more confidence could be assured in the precise tumor position and motion patterns. It is precisely the goal of this research to address this need.

The goal of this research, improving tumor motion predictions, has been used to guide the development of an appropriate physical model of the tracheobronchial system. Keeping this overall goal in mind has allowed for several assumptions to be made regarding the model physics, without any significant depreciation of overall motion prediction capabilities. First, the oral cavity and mouth have been removed from the flow models and the system inlet begins directly after the larynx. This assumption is justified by noting that the oral cavity flow cross sections are much larger than those within the trachea, therefore the pressure drop from the mouth to the start of the trachea will be insignificant when compared to the overall flow through the trachea. Our second physical simplification is to model the lung as a continuous porous medium with variable, patient specific, material properties. This approach has been used with success by previous investors for airflow modeling within lungs [3]. Our third and final simplification involves both physics and geometric assumptions. Since the trachea branches into smaller passageways numerous times before and within the lung, some decision must be made regrading where to separate the trachea from the lung itself. There are several trade-offs that must be investigated when making this decision, since as more levels are included: the flow passageways become increasing small so that eventually the continuum flow model assumptions will break down, the number of interfaces between the lung and trachea models increases, extracting the passageways from the medical images becomes more difficult, and consistency between the model and real patient is improved. We have found that the positive effect of increasing the number of branching levels on consistency is quickly degraded by the loss of geometric accuracy in obtaining the flow geometry from the medical images. Therefore, we have limited the branching to only the first two levels, as shown schematically in Figure 1.

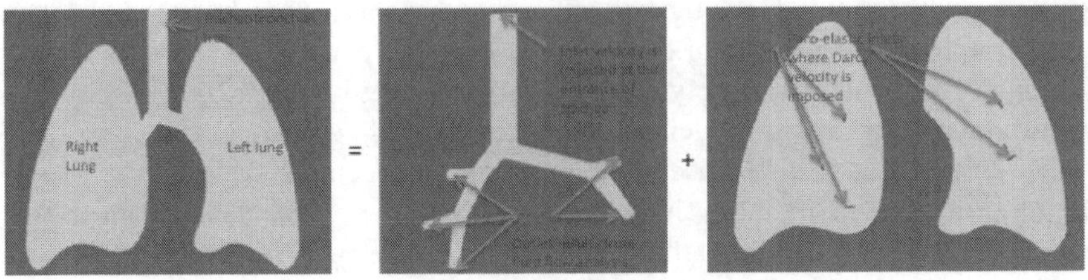


Figure 1: Coupling of the trachea and lung models.

Now that the major assumptions have been stated, we can begin to detail the overall process of modeling the tracheobronchial flow system. Mass flow rate and pressure data at the mouth, as well as computed tomography (CT) images are obtained for each patient during normal breathing. The CT images are then processed to obtain the flow geometries needed for computational modeling. The trachea is separated from the lungs at the second level of bifurcation so that two separate physical geometries are obtained. A Meshless incompressible flow modeling routine is then used to predict the airflow through the trachea over a normal breathing cycle, where the inlet conditions are obtained from the measured data. The outlets of the trachea flow model are connected to the lung model and provide the link between the lung and trachea flow fields. The incompressible flow modeling is carried out by solving the Navier-Stokes equations using the approach previously as described by Kassab and Divo [4].

The lungs are modeled as poro-elastic media, where the flow field satisfies Darcy's Law and the elasticity field is solved using Navier's equation:

$$\phi \beta_f \frac{\partial p}{\partial t} = \nabla \left[ \frac{\kappa}{\mu_f} (\nabla p) \right] - \frac{\partial}{\partial t} \left( \frac{\partial u_i}{\partial x_i} \right) \tag{1}$$

$$G \nabla^2 \vec{u} + \frac{G}{1 - 2\nu} \nabla (\nabla \cdot \vec{u}) = \nabla p \tag{2}$$

where $\vec{u}$ is the tissue deformation, $\nu$ is the constant tissue Poisson ratio, $\mu_f$ is the constant air viscosity, and $p$ is the pore pressure. The tissue properties in the above equations, porosity, $\phi$, permeability, $\kappa$, and shear modulus, $G$, are considered non-homogeneous as they can vary considerably throughout an individual's lung. The distribution of these properties for each patient are provided by a novel image processing technique as described by Santhanam et al. [2]. The overall coupling process is further illustrated using a simple 2D sketch in Figure 1. It should be noted that Eqn. (2) is given in steady-state form, while Eqn. (1) is transient. This is done as this is the approach utilized herein, as it is assumed that the lung deformation occurs at a much faster rate than the flow field changes within the porous lung tissue. Therefore, each time-step of the overall solution procedure requires the solution of the steady-state Navier equation, but much of the computational expense of this procedure is reduced since each time-step involves only a slight change in the deformation field solution.

## 3 Meshless modeling technique

The merits and benefits of Meshless Method solution approaches are well described in the literature, where it is generally noted that a reduction in the dependence of solution quality on the discretization is the most important achievement. Since this work relies on medical imaging data to provide the model geometries, it is difficult to guarantee that a smooth, clean geometry is obtained. This difficulty quickly leads one to assume that Meshless Methods are an excellent choice for use in this field of research, however, we have found that this is only true for certain formulations of Meshless Methods (the name Meshless Methods extends to include several distinct types of techniques). Furthermore, due to these geometric difficulties we have found that method robustness and stability are the key factors rather than accuracy in this field of numerical modeling. We have therefore adopted several different techniques, which are typically found within various Meshless Method approaches but rarely combined, into a single unified Meshless Method for the flow field modeling in this work. Specifically, this work utilizes radial basis function (RBF) interpolation, moving-least-squares (MLS) approximation, as well as finite differencing to create robust and efficient numerical solution schemes for both incompressible and poroelastic flows. Each of these techniques is applied in a localized manner to decrease computational effort and ensure the highest possible efficiency in the method. Some pertinent details regarding various Meshless-based methods as well each of the schemes mentioned are given in the remainder of this section, followed by an overview of how the various schemes are brought together to form a unified flow solution technique.

Typical Meshless Methods begin by assuming that any desired field, $\phi$, can be locally approximated by multiplying a set of basis functions, $\chi$, by a corresponding set of expansion coefficients, $\alpha$, as:

$$\phi(x) = \sum_{j=1}^{NF} \alpha_j \chi_j(x) \tag{3}$$

where $NF$ is a number of local expansion points (support domain) and $x$ refers to any number of coordinate directions. A critical component of these local Meshless techniques is the determination of a suitable set of basis functions that will accurately represent the field between the data points and provide appropriate estimations of the required field derivatives to facilitate the solution of the governing equation at hand. Many researchers, including the authors [4–7], have investigated this issue and several conclusions can be drawn from these previous works. For RBF methods, the Inverse Hardy Multiquadric [8] has repeatedly been noted as the best choice with regards to both stability and accuracy, while for MLS methods, standard monomial terms up to second order are typically found sufficient for most applications. Further details of these two approximation schemes are continued below.

Radial Basis Function interpolation has been the foundation of most Meshless Method formulations since the developments of Kansa over 20 years ago [9, 10]. Since those works, much more attention has been given to the specific details of the RBF schemes resulting in much gains in accuracy and stability of the techniques. Many researchers have reported such improved results by augmenting the RBFs with additional functions that are chosen specifically for the application at hand. Such augmentations are accomplished by updating the field approximation as:

$$\phi(x) = \sum_{j=1}^{NF} \alpha_j \chi_j(x) + \sum_{k=1}^{NP} \alpha_{(k+NF)} P_k(x) \tag{4}$$

where$P_k(x)$ refers to any desired set of basis functions and $NP$ refers to the number of extra basis functions used. Kassab and Divo [4], showed the importance of including constant and linear monomial terms, and Sarlar et al. [11] found much benefit to adding hyperbolic functions when solving heat transfer and other diffusion like equations. Despite the improvements gained by addition of these extra basis functions, much of the accuracy of the RBF techniques still depend heavily on the selection of a suitable “free parameter” that appears in the Hardy Multiquadric as:

$$\chi(x_i) = \frac{1}{\sqrt{r_j(x_i)^2 + c^2}} \tag{5}$$

where $x_i$ is the point at which the field is being estimated, $r_j$ is the Euclidean distance from $x_i$ to point a point $x_j$ within the local influence zone, and $c$ is the free parameter. Earlier works by the authors [4, 12] have presented the details of a method to optimize the value of this free parameter for each and every interpolation point in the field. This optimization has proven to provide a very reliable and accurate RBF interpolation scheme when using the Hardy Multiquadric RBF. Once the basis functions have been selected, the task of converting any desired partial differential equation into an algebraic expression becomes quite straight-forward, as each necessary differential operator may be applied directly to the basis functions and the resulting approximation equation may be solved by any standard type of explicit or implicit solution scheme.

Despite the excellent performance of RBF Meshless Methods and the relative ease of their implementation in many problems and application areas, they still suffer from three key deficiencies: convective derivatives are difficult to capture due to the radial symmetry of the RBFs; areas of locally Cartesian structure reduce to nearly the same results as standard finite differencing, but much more effort is expended in computing the required RBF weights; high gradient regions, such as shocks and boundary layers, can cause oscillations in the RBF interpolator, resulting in poor numerical instability. Therefore, other approximation schemes are necessary in some cases to alleviate these concerns.

Moving Least Squares approximations have gained a lot of attention in Meshless Methods in the area of global Meshless formulations [13–15], but have not seen much application in the area of localized approaches. However, our experience has found that MLS formulations provide much more stability when compared to RBF schemes, since much lower order approximation functions are utilized. Throughout our work we have found that quadratic MLS approximations are sufficiently accurate and very stable for a wide variety of application domains. For three-dimensional modeling, a quadratic MLS requires the following ten monomial terms:

$$\chi(x) = \left\{ 1 \quad x \quad y \quad z \quad xy \quad yz \quad zx \quad x^2 \quad y^2 \quad z^2 \right\} \tag{6}$$

It should be noted that since the MLS approximations in this work are applied only over a small influence domain and are fully independent from the approximations in other domains, additional weighting functions are not needed. Also, it should be noted when using MLS approximations the local support domain must be much larger than that of typical RBF interpolations, and our implementations have been found most promising when it is ensured that at least 20 points with a minimum of four unique values in each coordinate direction are included (in other words, the support domain nodes must lie in at least 4 distinct planes when viewed from each coordinate direction). The typical disadvantage of MLS schemes comes from the fact that they are approximating and not interpolating functions, therefore applying boundary conditions can be a very complex process. However, we have found that by combining the MLS approximation approach into a generalized finite differencing-like scheme, coined Virtual Finite Differencing in our previous work [12, 16], the application of both Dirichlet and Neumann type boundary conditions is straight-forward and obvious.

The above descriptions have been presented only with minor details as they are each well described in other works, however, what does not seem to have been undertaken by those working in the field of Meshless Methods is the combination of these techniques within a single computational implementation. Such a combination has been a major focal point of this research as the complex geometries present require much more stability than many other types of applications. We have attempted to optimize this combination through a rigorous testing and validation process.

Our recent works [16, 17] have shown that maintaining a Cartesian nodal structure throughout as much of the volume of the computational domain as possible allows classic finite differencing to be used in much of the domain (note that this is *exactly* equivalent to RBF Meshless methods for a 7-noded influence region as shown in previous works [12]), which significantly increases the computational efficiency. This approach of maintaining Cartesian nodal distributions has therefore been adopted in this work as well. This leaves only the boundary, near-boundary, and those interior regions where different discretization sizes come together where truly "Meshless" Methods must be used. For these interior nodes a nearly Cartesian structure still exists, which provides high quality local influence domains and allows RBF interpolation to be used without reservation. For the near boundary regions, the MLS Virtual Finite Differencing (MLS FD) approach is utilized, as stability is the major concern in this area. For the boundary itself, a combination approach exists. RBF interpolations are used directly on the boundary surface for the application of tangential boundary conditions, such as the traction conditions in elasticity applications. The MLS FD approach is then used for the normal direction boundary conditions or a pure finite difference is used when so-called "shadow nodes" are included in the computational domain (a thorough discussion of shadow nodes has been provided by the authors in several previous works [4, 12]). Lastly, when boundary derivative approximations are needed, but are not the applied boundary conditions (for example when computing the dilatation on the boundary in fluid applications) standard differentiation of the MLS approximating functions are utilized. The combination of schemes has been found to significantly improve the numerical stability of our Meshless Methods and further reduce the dependence of the solution on the quality of the underlying nodal distribution. It should be noted however, that as with any computational modeling scheme, accuracy is still a function of the local nodal spacing, therefore our implementations all include some type of automatic, solution based nodal refinement strategy, so that accuracy and "discretization convergence" is achieved for all our solutions.

## 4 Current results

The overall solution approach to tracheobronchial flow modeling utilized in this work has been verified for simulated two-dimensional configurations in our previous work [18]. These verifications provided some confidence to the soundness of the coupling schemes and the ability of the Meshless methods to capture the necessary flow physics. The extension to 3D and application to real patient lung geometries necessitated the solution approach updates outlined above. The approach is now capable of predicting lung tissue deformations for real patients and is currently undergoing verification in the clinic for real cancer patients.

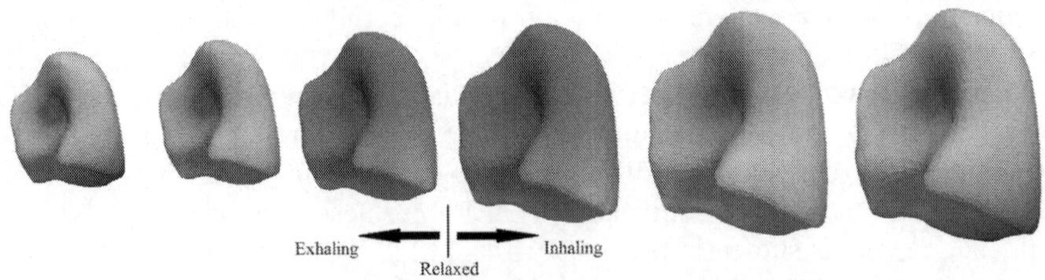


Figure 2: Example patient lung deformation results throughout breathing cycle (sized according to deformation and colored by deformation magnitude).

The lung deformation fields throughout time are shown below for one particular patient. Figure 2 shows the entire right lung and Figure 3 tracks the deformation history for a single point on the outer lung surface. An image reconstruction process is currently being completed that will allow the comparison of these deformation fields with the real patient lung motions obtained from time dependent (4D) CT scans. This process will enable the elasticity field boundary conditions to be properly varied with deformation, so that the computational models can more accurately match the real patient lung deformations. This next phase of research is currently underway for several patients.

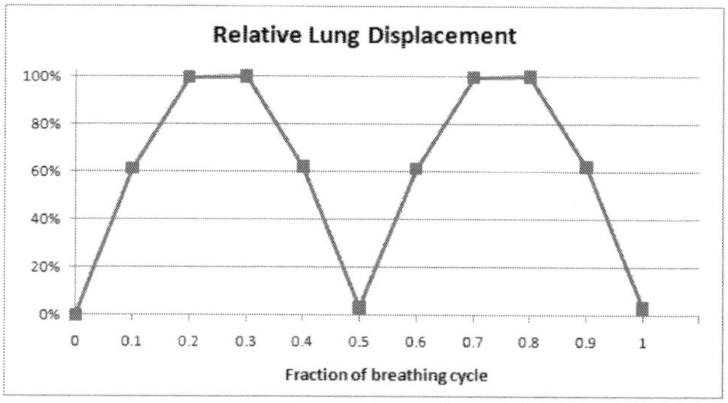


Figure 3: Relative displacement of a landmark point on the outer lung surface throughout the breathing cycle.

## 5 Conclusions

We have presented a new generalized Meshless Method approach that allows the tailoring of a Meshless method to achieve the most important goals for any application by selecting schemes with the most desirable properties for each different type of circumstance that is encountered within the solution process. It is suggested that: finite differencing be applied to the bulk of the interior to save effort on the pre-processing stages; RBF interpolation be applied along the boundary surfaces and within the volume for areas with suitable quality nodal discretizations; and the MLS approximations be used for near-boundary derivative estimations and for the application of generalized finite differencing in low quality distribution regions. Furthermore, it is our recommendation from this research that others explore such generalized approaches to improve other Meshless techniques in any areas of concern as well.

## Acknowledgement

We would like to acknowledge the funding received in support of this project from the James and Esther King Biomedical Research Program.

## References


[1] Stevens C, Munden R, Forster K, Kelly J, Zhongxing L, Starkschell G, Tucker S, Komaki R. Respiratory-driven lung tumor motions is independent of tumor size, tumor location, and pulmonary function. *International Journal of Radiation Oncology* 2001; **51**:62–68.

[2] Santhanam A, Willoughby T, Shah A, Meeks S, Rolland J, Kupelian P. Real-time simulation of 4d lung tumor radiotherapy using a breathing model. *MICCAI* 2008; **2**:710–717.

[3] Kowalczyk P. Mechanical model of lung parenchyma as a two-phase porous media. *Journal of Transport in Porous Media* 1993; **11**:281–295.

[4] Kassab A, Divo E. An efficient localized radial basis function meshless method for fluid flow and conjugate heat transfer. *ASME Journal of Heat Transfer* 2007; **129**:179–183.

[5] Divo E, Kassab A. An efficient localized rbf meshless method applied to fluid flow and conjugate heat transfer. *Proceedings of ASME IMECE*, 2005.

[6] Sarler B, Tran-Cong T, Chen C. Meshfree direct and indirect local radial basis function collocation formulations for transport phenomena. *Boundary Elements XVII*, Kassab A, Brebbia C, Divo E (eds.), WIT Press, 2005; 417–428.

[7] Shu C, Ding H, Yeo K. Local radial basis function-based differential quadrature method and its application to solve two-dimensional incompressible navier-stokes equations. *Computer Methods in Applied Mechanics and Engineering* 2003; **192**:941–954.

[8] Hardy RL. Multiquadric equations of topography and other irregular surfaces. *Journal of Geophysical Research* 1971; **76**:1905–1915.

[9] Kansa EJ. Multiquadrics: A scattered data approximation scheme with applications to computational fluid dynamics ii-solutions to parabolic, hyperbolic and elliptic partial differential equations. *Computers and Mathematics with Applications* 1990; **19**:147–161.

[10] Kansa EJ. Multiquadrics: A scattered data approximation scheme with applications to computational fluid dynamics i-surface approximations and partial derivative estimates. *Computers and Mathematics with Applications* 1990; **19**:127–145.

[11] Sarler B, Vertnik R. Local explicit radial basis function collocation method for diffusion problems. *Computers and Mathematics with Applications* 2006; **51**:1269–1282.

[12] Gerace S. An interactive framework for meshless methods analysis in computational mechanics and thermofluids. Master's Thesis, University of Central Florida 2007.

[13] Belytschko T, Lu YY, Gu L. Element-free galerkin methods. *International Journal of Numerical Methods* 1994; **37**:229–256.

[14] Belytschko T, Krongauz Y, Fleming M, Organ D, Liu WKS. Smoothing and accelerated computations in the element free galerkin method. *Journal of Computational and Applied Mathematics* 1996; **74**(1):111–126.

[15] Atluri SN, Zhu T. A new meshless local petrov-galerkin (mlpg) approach in computational mechanics. *Computational Mechanics* 1998; **22**:117–127.
[16] Gerace S, Erhart K, Divo E, Kassab A. Generalized finite difference meshless method in computational mechanics and thermofluids. *Proceedings of the Int. Conf. on Comp. Methods for Coupled Problems in Science and Engineering (COUPLED PROBLEMS 2009)*, 2009.
[17] Erhart K, Gerace S, Divo E, Kassab A. Turbulent compressible flow analysis with meshless methods. *Proceedings of the Int. Conf. on Comp. Methods for Coupled Problems in Science and Engineering (COUPLED PROBLEMS 2009)*, 2009.
[18] Huayamave V, Vidal A, Divo E, Kassab A, Santhanam A, Kupelian P. A meshless approach to solve the fluid poro-elastic interaction problem between the tracheo-bronchial tree and the lungs. *Proceedings of the Int. Conf. on Comp. Methods for Coupled Problems in Science and Engineering (COUPLED PROBLEMS 2009)*, 2009.

# Iterative meshfree methods for the Helmholtz equation

O. von Estorff & C. Wenterodt
*Institute of Modelling and Computation,*
*Hamburg University of Technology, Germany*

## Abstract

For acoustic computations in the mid-frequency range the finite element method (FEM) is a well-known standard tool. Unfortunately, for increasing frequencies, i.e. higher wavenumbers, the FEM suffers from the so-called pollution effect which is mainly a consequence of the dispersion, meaning that the numerical wavenumber and the exact wavenumber disagree. Using meshfree methods as, e.g., the radial point interpolation method (RPIM) or the element-free Galerkin method (EFGM) can reduce this effect significantly.

Moreover, meshfree methods allow the usage of shape functions that can be adapted to the differential equation to be solved. Consequently, an iterative method can be derived, which uses a standard meshfree method to compute a first approximation for the given problem. In a second step this approximation is taken to construct new shape functions that are able to better reproduce the wave-like character of the solution. If a few requirements are considered, this method leads to better results in a more efficient way.

In this paper two meshfree methods, namely an iterative RPIM and an iterative EFGM, are examined. The methods are compared to the FEM and restrictions for an efficient applicability are shown.
*Keywords: Helmholtz equation, meshfree methods, acoustics, iterative method, EFGM, RPIM.*

## 1 Introduction

In recent years meshfree methods have attracted more and more attention in several disciplines of computational mechanics [1]. Especially in the area of computational acoustics, it is well-known that the standard methods as, e.g., the finite

WIT Transactions on Modelling and Simulation, Vol 50, © 2010 WIT Press
www.witpress.com, ISSN 1743-355X (on-line)
doi:10.2495/BE100141

element method (FEM) or the boundary element method (BEM) possess some drawbacks. The FEM suffers from the pollution effect. Especially for high wavenumbers the results become inaccurate since the dispersion error increases [2]. For the BEM an increasing number of elements leads to a significantly longer computation time, since the system matrices are fully populated and possibly non-symmetric [3].

In contrast, meshfree methods offer some advantages. For the construction of the shape functions no mesh is needed. Instead, shape functions are constructed with respect to field points which can be arbitrarily distributed inside the domain under consideration. Amongst others, this makes the implementation of adaptive methods rather straightforward.

With respect to acoustic computations, in comparison to the FEM the dispersion effect can be drastically reduced for the element free Galerkin method (EFGM) [4] as well as for the radial point interpolation method (RPIM) [5]. Furthermore, the shape functions of the meshfree methods can be easily adapted to the problem to be solved. Consequently, Lacroix et al. [6] introduced an iterative EFGM which is based on problem-dependent shape functions. In a first step, an approximation for the sound pressure distribution inside the given domain is computed using a standard EFGM. From this solution the phase distribution can be obtained and then be used for the construction of new phase-dependent shape functions. These shape functions are better suited to reproduce the wave-like character of the solution and thus the results can be improved.

In this paper the method of Lacroix et al. has been further examined. In fact, own investigations have shown, that the accuracy of the iterative EFGM is highly dependent on the numerical integration scheme. Furthermore, different weight functions have been introduced and compared in order to optimize the method. Additionally, an iterative RPIM has been implemented; it will be compared to the iterative EFGM.

## 2 Acoustics

The propagation of pressure waves in an acoustic fluid inside a domain $\Omega$ is governed by the wave equation which can be derived using the balance of mass and momentum as well as the ideal gas law, provided that the state variables pressure $P$, density $\rho$, and velocity $v$ experience only small variations. For time harmonic waves of frequency $\omega$ the sound pressure $p$ is the solution of the Helmholtz equation

$$\nabla^2 p + k^2 p = 0 \tag{1}$$

with wavenumber $k = \omega/c$ and speed of sound $c$. The boundary $\Gamma$ of the fluid domain $\Omega$ is decomposed into three distinct regions $\Gamma_D$, $\Gamma_N$, and $\Gamma_R$. The associated boundary conditions are

- Dirichlet boundary conditions on $\Gamma_D$:

$$p = \bar{p} \tag{2}$$

- Neumann boundary conditions on $\Gamma_N$:

$$\frac{\partial p}{\partial \mathbf{n}} = -i\omega\rho\bar{v}_n \tag{3}$$

- Robin boundary conditions on $\Gamma_R$:

$$\frac{\partial p}{\partial \mathbf{n}} = -i\omega\rho A_n p \tag{4}$$

Here, $A_n$ represents the admittance, $\mathbf{n}$ is the normal vector and $v_n$ the normal velocity.

# 3 Meshfree methods

In comparison to mesh-based methods as, e.g., the FEM, meshfree methods do not need a mesh for the representation of the domain $\Omega$ to be examined. In $\Omega$ and on its boundary $\Gamma$, field points $\mathbf{x}_i$ can be arbitrarily distributed. An approximation of the sound pressure $p^h$ at an arbitrary point $\mathbf{x}_Q$ in $\Omega$ is obtained by interpolation of the form

$$p^h(\mathbf{x}_Q) = \sum_{i=1}^{n} \Phi_i(\mathbf{x}_Q) p_i. \tag{5}$$

Here, $p_i$ is the coefficient of the solution vector $\mathbf{p}$ belonging to field point $\mathbf{x}_i$ and $\Phi_i$ is the associated shape function which is constructed using all $n$ field points that influence $\mathbf{x}_Q$. In order to obtain an equation system that can be solved for the coefficient vector $\mathbf{p}$, the shape functions are substituted into the weak form of the Helmholtz equation both as test and as trial functions. The resulting equations have to be integrated numerically. For this reason a background mesh has to be introduced, which can be chosen independently of the field point distribution and which does not have to represent the domain $\Omega$.

# 4 Shape functions

## 4.1 EFGM shape functions

In order to obtain the shape functions of the EFGM, the unknown sound pressure $p^h$ is interpolated by

$$p^h(\mathbf{x}) = \sum_{i=1}^{n} P_i(\mathbf{x}) a_i(\mathbf{x}) = \mathbf{P}^T(\mathbf{x})\mathbf{a}(\mathbf{x}). \tag{6}$$

Here, $\mathbf{P}^T(\mathbf{x})$ denotes a basis, which can be chosen as e.g.

$$\mathbf{P}^T(\mathbf{x}) = [1 \quad x \quad y] \tag{7}$$

in the two-dimensional case. The coefficient vector $\mathbf{a}(\mathbf{x})$ is determined by minimization of a functional such that the approximation $p^h(\mathbf{x}_i)$ at the field points yields the best possible agreement with the according coefficients $p_i$ of the solution vector. This leads to

$$\mathbf{a}(\mathbf{x}) = \mathbf{A}^{-1}(\mathbf{x})\mathbf{B}(\mathbf{x})\mathbf{p} \tag{8}$$

with

$$\mathbf{A}(\mathbf{x}) = \sum_{i=1}^{n} w_i(\mathbf{x})\mathbf{P}(\mathbf{x}_i)\mathbf{P}^T(\mathbf{x}_i) \tag{9}$$

$$\mathbf{B}(\mathbf{x}) = [w_1(\mathbf{x})\mathbf{P}(\mathbf{x}_i), \ldots, w_n(\mathbf{x})\mathbf{P}(\mathbf{x}_n)]\,. \tag{10}$$

In eqns (9) and (10) the weight functions $w_i(\mathbf{x})$ can be chosen, e.g., as the cubic spline

$$w_i(r) = \begin{cases} \frac{2}{3} - 4r^2 + 4r^3 & \text{for } r \leq \frac{1}{2} \\ \frac{4}{3} - 4r + 4r^2 - \frac{4}{3}r^3 & \text{for } \frac{1}{2} < r \leq 1 \\ 0 & \text{for } r > 1 \end{cases} \tag{11}$$

the quartic spline

$$w_i(r) = \begin{cases} 1 - 6r^2 + 8r^3 - 3r^4 & \text{for } r \leq 1 \\ 0 & \text{for } r > 1 \end{cases} \tag{12}$$

or the exponential function

$$w_i(r) = \begin{cases} \dfrac{\exp(-(2r)^2) - \exp(-4)}{1 - \exp(-4)} & \text{for } r \leq 1 \\ 0 & \text{for } r > 1 \end{cases} \tag{13}$$

with $r$ denoting the distance between point $\mathbf{x}$ and field point $\mathbf{x}_i$ scaled by the influence radius $r_{infl}$, i.e.

$$r = \frac{||\mathbf{x} - \mathbf{x}_i||}{r_{infl}}. \tag{14}$$

Substituting eqn (8) into eqn (6) yields

$$p^h(\mathbf{x}) = \mathbf{P}^T(\mathbf{x})\mathbf{A}^{-1}(\mathbf{x})\mathbf{B}(\mathbf{x})\mathbf{p} = \mathbf{\Phi}(\mathbf{x}) \cdot \mathbf{p} \tag{15}$$

with the vector of shape functions

$$\mathbf{\Phi}(\mathbf{x}) = \mathbf{P}^T(\mathbf{x})\mathbf{A}^{-1}(\mathbf{x})\mathbf{B}(\mathbf{x}). \tag{16}$$

### 4.2 RPIM shape functions

The shape functions for the RPIM can be obtained by an ansatz of the form

$$\begin{aligned} p^h(\mathbf{x}, \mathbf{x}_Q) &= \sum_{i=1}^{n} R_i(\mathbf{x}) a_i(\mathbf{x}_Q) + \sum_{j=1}^{n} P_j(\mathbf{x}) b_j(\mathbf{x}_Q) \\ &= \mathbf{R}^T(\mathbf{x})\mathbf{a}(\mathbf{x}_Q) + \mathbf{P}^T(\mathbf{x})\mathbf{b}(\mathbf{x}_Q). \end{aligned} \tag{17}$$

Here, $R_i$ is a radial basis function which can be chosen as, e.g., the multiquadrics

$$R_i(r) = (r^2 + (\alpha_c d_c)^2)^q \tag{18}$$

where $r$ denotes the distance between point $\mathbf{x}$ and field point $\mathbf{x}_i$. The characteristic length $d_c$ is the average field point spacing in the influence domain and $\alpha_c$ and $d_c$ can be adapted to the problem. The additional polynomial basis $\mathbf{P}^T(\mathbf{x})$ in eqn (17) ensures the consistency of the method and can be chosen as for the EFGM in eqn (7). The unknown coefficients $a_i$ and $b_j$ are chosen such that eqn (17) is fulfilled at all $n$ field points that influence point $\mathbf{x}_Q$.

It should be noted that the RPIM shape functions possess the Kronecker delta property, which is not the case for the EFGM shape functions. The coefficient vector $\mathbf{p}$ contains sound pressure values at the field points only if the method fulfills the Kronecker delta property. More information about the construction of meshfree shape functions and their properties can be found in [1].

## 5 Iterative meshfree method

Meshfree shape functions of the RPIM and the EFGM can be easily adapted to the problem to be solved. This concept has, e.g., been used for the examination of crack tip fields [7]. For acoustic problems that are described by the Helmholtz equation, the sound pressure at a point $\mathbf{x}$ can be written as

$$p(\mathbf{x}) = \hat{P}(\mathbf{x}) \left(\cos\theta(\mathbf{x}) + \mathrm{i}\sin\theta(\mathbf{x})\right) \tag{19}$$

with $\hat{P}(\mathbf{x})$ being the amplitude and $\theta(\mathbf{x})$ being the phase of the wave. The polynomial basis in eqn (7) or (17) is now substituted by the basis

$$\mathbf{P}^T(\mathbf{x}) = [1 \quad \cos\theta(x,y) \quad \sin\theta(x,y)]\,. \tag{20}$$

Provided that the phase is exactly known in the whole domain $\Omega$, the shape functions constructed with the basis (20) can interpolate the sound pressure exactly. Since the phase is generally unknown, a first approximation of the sound pressure $p_1^h(\mathbf{x})$ is computed using a standard EFGM or RPIM, respectively. From this pressure field an approximation of the phase distribution can be obtained and then be used for the construction of the adapted shape functions. Using the adapted shape functions, a second approximation $p_2^h(\mathbf{x})$ is computed. This algorithm can either be repeated for a predefined number of iterations or aborted if a convergence criterion is reached.

The method described above has been proposed and tested by Lacroix et al. [6] for the EFGM. Own investigations have shown that the success of the iterative method is highly dependent on the quality of the numerical integration. Furthermore the methodology can be optimized if proper weight functions are used. Moreover, the applicability of phase-dependent RPIM shape functions has not yet been investigated.

# 6 Numerical investigations

In the following, the iterative methods as well as the FEM are applied to an acoustic problem that can also be solved analytically. The square domain $\Omega$ shown in figure 1 with edge length $L = 1\,\mathrm{m}$ is investigated. At $x = y = 0$ a Dirichlet boundary condition with $p_0 = 1\,\mathrm{N/m^2}$ is prescribed. On all four edges Robin boundary conditions are chosen such that the analytical solution is a plane wave of the form

$$p(x, y) = \cos(k(x\cos\beta + y\sin\beta)) + \mathrm{i}\sin(k(x\cos\beta + y\sin\beta)). \qquad (21)$$

Here, $\beta$ denotes the propagation angle and is defined as $\beta = 20°$ in the following. The fluid properties are chosen as density $\rho = 1.247\,\mathrm{kg/m^3}$ and speed of sound $c = 343\,\mathrm{m/s}$ which correspond to air.

Figure 2 shows the $L_2$ error of the iterative methods for a wavenumber $k = 1\,\frac{1}{\mathrm{m}}$ and field point distance $h_{fp} = 0.1\,\mathrm{m}$. The background mesh is chosen in accordance to the field point distribution, i.e. the size of the background cells is $h_{bg} = 0.1\,\mathrm{m}$. The number of gauss points in each cell is $n_{qp} = 10 \times 10$. Additionally, the error of the FEM is shown. For the FEM no iterative method has been implemented and therefore the error of the FEM is constant here.

It is obvious that the EFGM yields the best result. The first iteration which is carried out using the standard EFGM already leads to an error which is only $0.4\%$

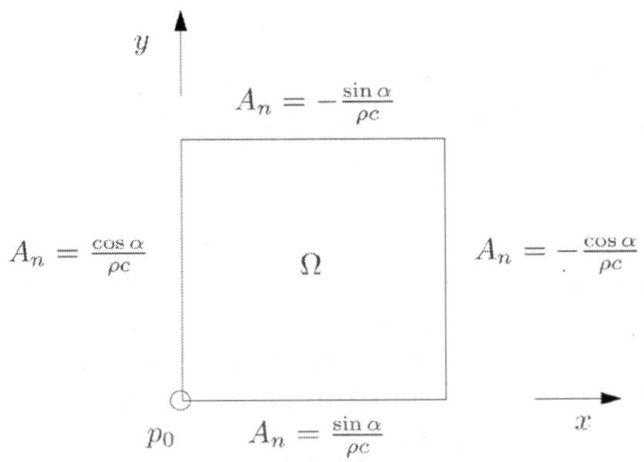


Figure 1: Acoustic domain $\Omega$ with Robin and Dirichlet boundary conditions.

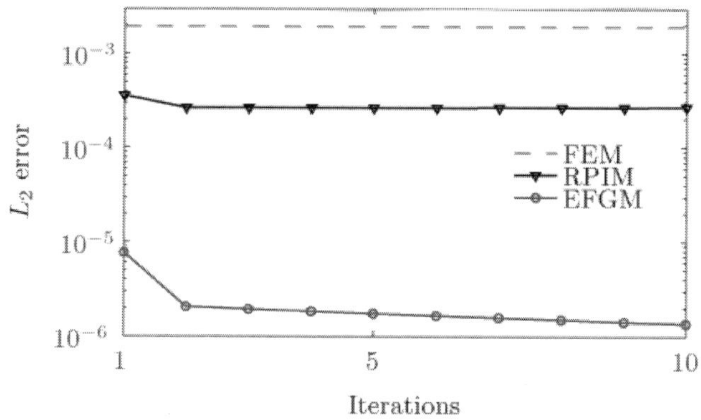


Figure 2: $L_2$ error of the RPIM, EFGM, and FEM for $k = 1\,\frac{1}{\mathrm{m}}$ and $h_{fp} = 0.1\,\mathrm{m}$.

of the FEM error. The RPIM leads to at least $18\%$ of the FEM error. During the following iterations the error of the EFGM and the RPIM can be further reduced, although the effect is much clearer for the EFGM.

In order to show the local effect of the iterative EFGM figure 3(a) contains the error distribution on $\Omega$ for the first iteration which corresponds to the standard EFGM. Furthermore in figure 3(b) the error distribution after the second iteration is shown. The error could be significantly reduced. For comparison the error of an optimal EFGM is presented in figure 3(c). Here, "optimal EFGM" means, that the shape functions are constructed using the phase distribution of the analytical solution. As already mentioned before, the resulting shape functions are dispersion free and are able to represent the solution exactly. The error of the optimal EFGM almost vanishes but it is not zero everywhere. The reasons for this will be explained later.

Several computations have shown, that the capabilities of the iterative EFGM are highly dependent on the quality of the numerical integration. For the standard EFGM the numerical integration does not have a significant influence on the results, but the phase-dependent shape functions appear to be very sensitive to the choice of the background mesh and the numerical integration order. Figure 4 shows the $L_2$ error of the EFGM for different background meshes. The size of the background cells $h_{bg}$ as well as the number of integration points in each cell $n_{qp}$ have been varied. If a coarse background mesh with a low integration order (in this case $h_{bg} = 0.1\,\mathrm{m}$ and $n_{qp} = 4 \times 4$) is chosen, the iterative method even deteriorates the results of the standard EFGM in the following iterations.

Similar investigations for the RPIM have shown that the background mesh and integration order do not influence the quality of the method. On the other hand, the results of the RPIM cannot keep up with the EFGM. For this reason the following analysis is carried out for the EFGM only.

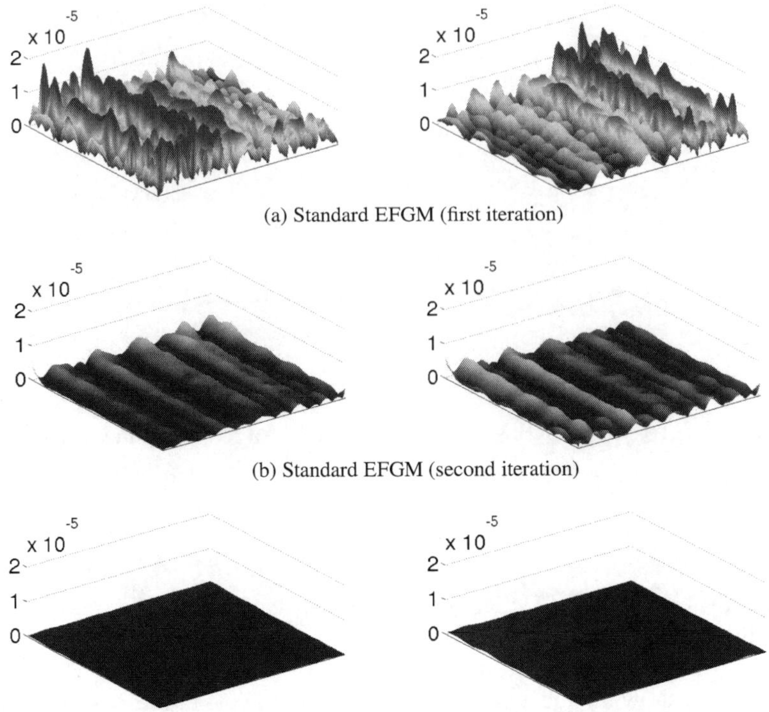


(a) Standard EFGM (first iteration)

(b) Standard EFGM (second iteration)

(c) Optimal EFGM

Figure 3: Error distribution of the standard, iterative, and optimal EFGM for $k = 1\,\frac{1}{\mathrm{m}}$, $h_{fp} = 0.1\,\mathrm{m}$, $h_{bg} = 0.1\,\mathrm{m}$, and $n_{qp} = 10 \times 10$ (left: real part, right: imaginary part of the error).

Up to now, only 10 iterations have been considered. Figure 5 shows the results of the iterative EFGM (EFGM iterative) for 10 as well as 100 iterations. For comparison, the $L_2$ error of the optimal EFGM (EFGM opt) is plotted. This means that for the first iteration the optimal EFGM with shape functions depending on the analytical phase distribution has been used and in the subsequent iterations the shape functions have been computed using the phase of the solution obtained from the last iteration.

The optimal EFGM reaches an $L_2$ error of $4 \cdot 10^{-7}$ in the first iteration. Due to numerical instabilities this value oscillates slightly during the next iterations. The iterative EFGM converges towards a limit of $7 \cdot 10^{-7}$.

In order to assess the influence of the numerical integration on the results in a more objective way, the optimal EFGM is further examined. In theory, the optimal EFGM should yield an optimal solution because the shape functions interpolate the sound pressure exactly. Nevertheless, the solution is not exact due to errors of the numerical integration and other instabilities that result from e.g. matrix inversions. Figure 6 shows the $L_2$ error of the optimal EFGM for different sizes $h_{bg}$ of the

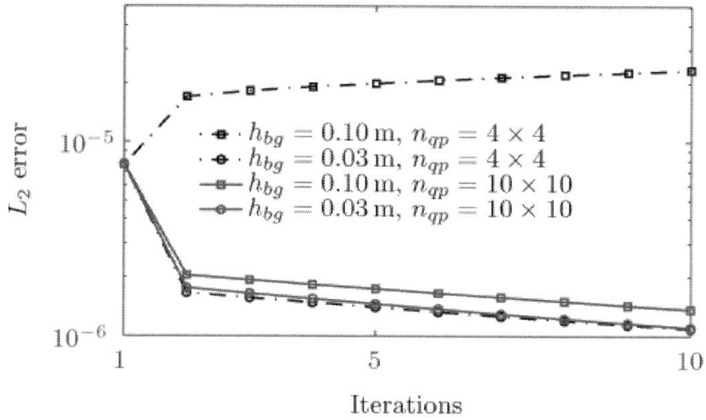


Figure 4: $L_2$ error for different background meshes for $k = 1\,\frac{1}{\mathrm{m}}$ and $h_{fp} = 0.1\,\mathrm{m}$.

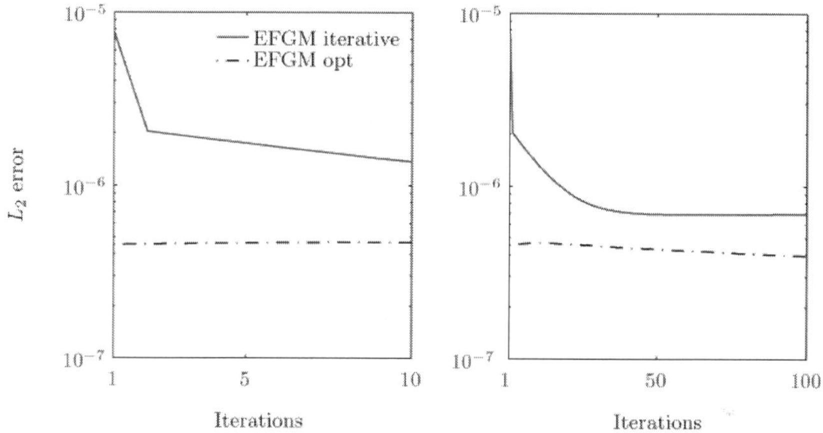


Figure 5: $L_2$ error of the EFGM and the optimal EFGM for $k = 1\,\frac{1}{\mathrm{m}}$, $h_{fp} = h_{bg} = 0.1\,\mathrm{m}$, and $n_{qp} = 10 \times 10$.

background cells (left) and different numbers of gauss points $n_{qp}$ (right). The influence of the integration scheme can be clearly seen: If the cells are chosen too coarse or if the integration order is too low, the optimal EFGM can only yield an $L_2$ error of about $10^{-4}$ to $10^{-3}$. In contrast, the $L_2$ error can be reduced to less than $10^{-8}$ if the integration is improved. Furthermore, the method leads to better results if the size of the background cells is a multiple of the field point spacing and vice versa.

Finally, the influence of different weight functions $w_i(\mathbf{x})$ on the results of the EFGM are examined. Up to now, all computations have been performed using the

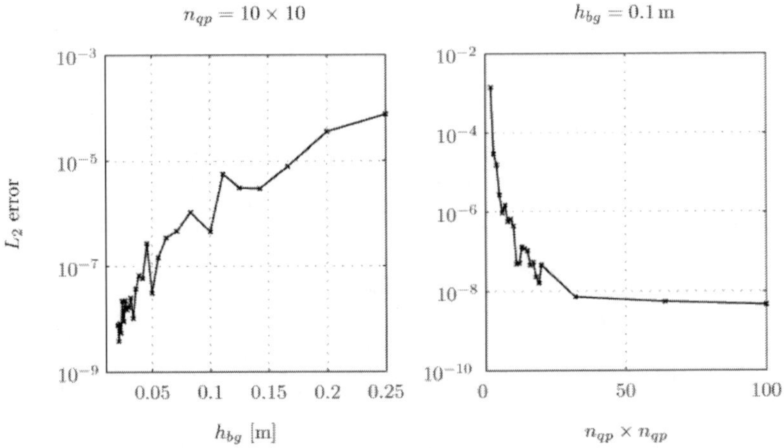


Figure 6: $L_2$ error of the optimal EFGM ($k = 1\,\frac{1}{\text{m}}$, $h_{fp} = 0.1\,\text{m}$).

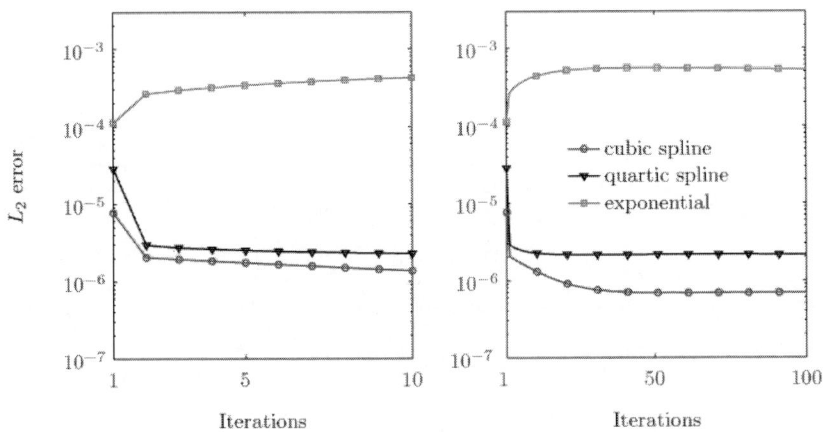


Figure 7: $L_2$ error of the EFGM for different weight functions for $k = 1\,\frac{1}{\text{m}}$, $h_{fp} = h_{bg} = 0.1\,\text{m}$, and $n_{qp} = 10 \times 10$.

cubic spline. In [6] Lacroix et al. proposed exponential weight functions. Figure 7 shows the $L_2$ error of the iterative EFGM if the cubic spline in eqn (11), the quartic spline in eqn (12) and the exponential function in eqn (13) are used. Obviously the weight function has a considerable influence on the error of not only the standard EFGM (first iteration) but also of the iterative EFGM. In the considered case the cubic spline weight functions yield the lowest error. In contrast, the iterative EFGM with exponential weight function even increases the error of the standard EFGM.

## 7 Conclusions

In this contribution, two meshfree methods, namely the EFGM and the RPIM, have been applied to a two-dimensional acoustic problem. An iterative method based on shape functions that are adapted to the problem has been implemented and investigated especially with respect to the numerical integration scheme. Furthermore, for the EFGM, the influence of different weight functions has been examined.

The EFGM as well as the RPIM lead to a lower $L_2$ error than the FEM for similar discretizations. The iterative methods further reduce this error, but only if the numerical integration scheme is chosen appropriately. The influence of the background mesh and integration order on the iterative method is less considerable for the RPIM. On the other hand, the EFGM leads to much lower errors than the RPIM.

Furthermore, the influence of different weight functions on the quality of the solution has been examined. For the considered example, the cubic spline performed best whereas the exponential weight function did not yield reasonable results.

In order to be able to use the iterative methods in practice, further investigations are necessary, e.g., application to real life problems and the analysis of the computation time.

## References

[1] Liu, G., *Mesh Free Methods: Moving beyond the Finite Element Method*. CRC Press: Boca Raton, 2003.

[2] Deraemaeker, A., Babuška, I. & Bouillard, P., Dispersion and pollution of the FEM solution for the Helmholtz equation in one, two and three dimensions. *International Journal for Numerical Methods in Engineering*, **46(4)**, pp. 471–499, 1999.

[3] von Estorff, O., (ed.) *Boundary Elements in Acoustics: Advances and Applications*. WIT Press: Southampton, 2000.

[4] Suleau, S., Deraemaeker, A. & Bouillard, P., Dispersion and pollution of meshless solutions for the Helmholtz equation. *Computer Methods in Applied Mechanics and Engineering*, **190(5)**, pp. 639–657, 2000.

[5] Wenterodt, C. & von Estorff, O., Dispersion analysis of the meshfree radial point interpolation method for the Helmholtz equation. *International Journal for Numerical Methods in Engineering*, **77(12)**, pp. 1670–1689, 2009.

[6] Lacroix, V., Bouillard, P. & Villon, P., An iterative defect-correction type meshless method for acoustics. *International Journal for Numerical Methods in Engineering*, **57(15)**, pp. 2131–2146, 2003.

[7] Fleming, M., Chu, Y., Moran, B. & Belytschko, T., Enriched element-free Galerkin methods for crack tip fields. *International Journal for Numerical Methods in Engineering*, **40**, pp. 1483–1504, 1997.

# Control volume-radial basis function method for two-dimensional non-linear heat conduction and convection problems

C. A. Bustamante[1], W. F. Flórez[1], H. Power[2], M. Giraldo[1] & A. F. Hill[1]
[1]*Institute of Energy, Materials and Environment, Universidad Pontificia Bolivariana, Colombia*
[2]*School of Mechanical, Materials and Manufacturing Engineering, University of Nottingham, UK*

## Abstract

An improvement to the traditional Finite Volume Method (FVM) for the solution of boundary value problems is presented. The new method applies the local Hermitian interpolation with Radial Basis Functions (RBF) as an interpolation scheme to the FVM discretization. This approach, called the Control Volume-Radial Basis Function (CV-RBF) method, uses an interpolation scheme based on the meshless Symmetric method, in which the numerical solution is approximated by employing the governing equation and the boundary condition operators. The RBF implemented is the Multiquadric (MQ) with a shape parameter found experimentally. The two-dimensional solutions to the Dirichlet problem for linear heat conduction, heat transfer in the Poiseuille flow and the non-linear conduction situations are obtained by the CV-RBF method. The numerical results are in agreement with the corresponding analytical and numerical solutions found in the literature.
*Keywords: radial basis function, finite volume method, heat transfer, local Hermitian interpolation.*

## 1 Introduction

The Finite Volume Method (FVM) has become one of the preferred discretization strategies in the field of transport phenomena. The interpolation scheme used to

WIT Transactions on Modelling and Simulation, Vol 50, © 2010 WIT Press
www.witpress.com, ISSN 1743-355X (on-line)
doi:10.2495/BE100151

approximate the diffusive and convective fluxes on the volume faces is an important aspect for improving the FVM. The compilation of interpolation schemes done by Patankar [1] and later by Versteeg and Malalasekera [2] and Wesseling [3], has been enlarged to a variety of strategies with the aim of decreasing the truncation error preserving convergence, stability and consistency. In addition, interpolation schemes that are able to treat unstructured meshes are necessary to get solutions for the transport phenomena equations in domains with complex geometry.

To achieve an efficient and accurate approximation of the fluxes at the volume faces on unstructured meshes, several authors proposed schemes that use traditional FVM in conjunction with high order approximation methods, for instance, a two-dimensional polynomic approximation [4], the Least Square Reconstruction Technique (LSRT) [5] and the Gauss-Green Reconstruction technique (GGRT) [6, 7]. In addition, the geometric treatment required to compute approximated fluxes is the cause of the high dependence of the FVM on the mesh configuration. A meshless interpolation scheme can be a good option to weaken this relationship. According to the above, the Radial Basis Function (RBF) collocation method is used here as an interpolation scheme, which results in a high order approximation [8] that is able to fit quasi-scattered data.

Currently the RBF collocation is widely used in the scattered data interpolation. In addition, it is the base of several meshless methods used to solve Partial Differential Equations (PDEs). The first attempt to develop a meshless method by RBF collocation was made by Kansa [9]. Kansa used the Multiquadric (MQ) function to obtain an accurate meshless solution to the advection-diffusion and Poisson equations without employing any special treatment for the advection term (upwinding), due to the high order of the resultant scheme and the intrinsic relationship between governing equations and the interpolation. This strategy involved all the nodes in the domain, therefore it produced a global full matrix. Although the Kansa's method was used by many authors to solve the Laplace, the Poisson, the Helmotz and Parabolic equations, showing better accuracy compared to traditional methods [10–13], Kansa and Hon [14] concluded that the matrix ill-conditioning worsens as the number of nodes increases. In the case of using the MQ function, a singular matrix is obtained for some values of the shape parameter. Fasshauer [15] used the Hermite interpolation to construct an approximated function that resulted in a symmetric matrix and therefore a lesser amount of computation. He concluded that, for the PDEs solved, the Hermite interpolation or the Symmetric method perform slightly better than the Kansa or Unsymmetric methods. Jumarhon et al. [16] obtained a similar improvement using the Symmetric method and Power and Barraco [17] obtained better results by employing the Symmetric method for a variety of problems, including the convection diffusion equation.

Due to the ill-conditioned matrix of the global RBF collocation, it was implemented locally to improve the traditional methods for solving PDEs. The Finite Difference Method (FDM) was modified by Wright and Fornberg [8] to obtain higher order approximation by employing RBF Hermite interpolation. In the Finite

Volume Method (FVM), the RBF collocation was first implemented as an interpolation scheme by Moroney and Turner [18] to solve the two-dimensional nonlinear anisotropic diffusion equation and later, by Moroney and Turner [19] for the three-dimensional case in the same situation. Recently Orsini et al. [20] solved the diffusion convection equation by the FVM in conjunction with the RBF Hermite interpolation scheme. They had the convective and diffusive fluxes in terms of neighboring node values according to the Symmetric method and called this strategy the Control Volume-Radial Basis Function (CV-RBF).

In this work, the CV-RBF method is used to solve the two-dimensional heat equation for steady situations. In the next section the mathematical model that governs the steady heat convection and conduction is presented. After, the CV-RBF method is explained and the final discrete equation for the generic convection-diffusion problem is obtained. Finally, the numerical result are presented and compared with numerical and analytical solutions found in literature.

## 2 Governing equations

From the energy balance in an incompressible continuum medium, the heat equation (1) is obtained by subtracting the mechanical energy, expressing the internal energy in terms of the temperature and neglecting the accumulation and viscous dissipation terms. In this expression $\rho$ is the medium density, $\hat{C}_p$ is the specific heat at constant pressure, $\vec{v}$ is the velocity field and $S$ is the source term, which can depend on temperature and space.

$$\rho \hat{C}_p v_j \frac{\partial T}{\partial x_j} = -\frac{\partial q_j}{\partial x_j} + S(T, \vec{x}) \tag{1}$$

The molecular phenomena responsible of the conduction heat flux $\vec{q}$ in an isotropic system can be quantified on a macroscopic level by Fourier's Law (2), where $k$ is the thermal conductivity.

$$q_i = -k(T)\frac{\partial T}{\partial x_i} \tag{2}$$

## 3 Control volume-radial basis function

The CV-RBF method improves the traditional FVM by an interpolation scheme based on the Hermitian RBF. In this section, the formulation of the CV-RBF method developed by Orsini et al. [20] applied to the heat equation is presented.

### 3.1 Finite Volume Method discretization

The discretization process is applied to the heat equation (1) after substituting the Fourier's law (2) in it. First, the governing equation is integrated over a generic volume and the divergence theorem is applied to obtained the expression (3), where

$\Omega$ and $\partial\Omega$ are the finite volume and its surrounding surface, respectively.

$$\oint_{\partial\Omega} k\frac{\partial T}{\partial x_i} n_i d\partial\Omega - \oint_{\partial\Omega} \rho\hat{C}_p v_i T n_i d\partial\Omega = -\int_{\Omega} S d\Omega \qquad (3)$$

Using non-overlapping polygonal volumes, the surface integral can be expressed as the sum of the integrals over each of the $m$ volume faces. Then, the surface and volume integrals have to be approximated by a numerical method. In this case for simplicity and according to traditional FVM, the mid-point rule is employed both for the surface and volume integrals. The resultant expression (4) is the final FVM discretized equation. The volume and surface geometric centers are $\vec{x}_p$ and $\vec{x}_l$, respectively.

$$\sum_{l=1}^{m}\left(k\frac{\partial T}{\partial x_i} n_i \Delta\partial\Omega\right)\Bigg|_{\vec{x}=\vec{x_l}} - \sum_{l=1}^{m}\left(\rho\hat{C}_p v_i T n_i \Delta\partial\Omega\right)|_{\vec{x}=\vec{x_l}} = -S(\vec{x}_p)\Delta\Omega \qquad (4)$$

To complete the solution procedure, the dependent variable and its gradient, at the surface central points, have to be approximated in terms of the neighbor volume central nodes according to an interpolation scheme.

### 3.2 Local Hermitian Interpolation scheme

The Hermitian Interpolation scheme is a symmetrical RBF collocation technique used to solve PDEs. The RBF main feature is the only dependence in Euclidian distance between the evaluation point $\vec{x}$ and the trial nodes $\vec{\xi}$. Another characteristic is the monotonically increasing or decreasing of the function with the distance from the evaluation point. Here, the RBF used is the Multiquadric MQ function (5), where $r = \|\vec{x} - \vec{\xi}\|$ is the Euclidian distance and $c$ is the shape parameter which allows to modified the slope of the function. The MQ function converges exponentially and is conditionally positive defined of order $m > 0$, since a $m - 1$ order polynomial has to be added to get a non-singular interpolation matrix. In this case, $m = 1$ and the shape parameter is obtained by numerical experiments.

$$\Psi(r) = (r^2 + c^2)^{m/2} \qquad (5)$$

A boundary value problem is considered for the dependent variable $\phi$. The problem is defined by equations (6) and (7), where $L$ and $B$ are linear differential operators that apply over the domain $\Omega$ and the boundary $\partial\Omega$, respectively.

$$L[\phi(\vec{x})] = f(\vec{x}) \qquad (6)$$

$$B[\phi(\vec{x})] = g(\vec{x}) \qquad (7)$$

In RBF collocation strategies to solve PDEs are two basic procedures: Kansa's Method [9] and Symmetric Method [21]. The Hermite Interpolation scheme is used on the Symmetric method. A variation of the latter Symmetric method was developed by Orsini et al. [20], in which the Hermite Interpolation applies locally.

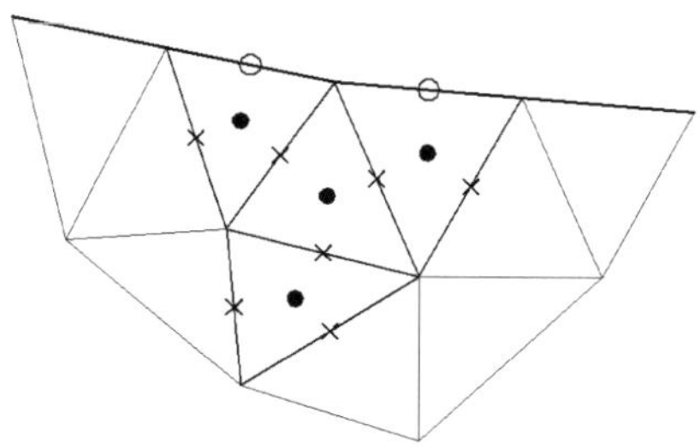

Figure 1: Node locations inside the stencil: • central nodes (NP), ∘ boundary points (NB), × internal nodes (NPDE).

In this case, it is necessary to define a subdomain which contains the nodes. In the CV-RBF method, the subdomain is called stencil and it contains the central volume and at least one neighbor volume. The stencil includes $NP$ central nodes or points where the variable is unknown, $NB$ boundary points where the boundary operator is applied and $NPDE$ internal locations where the governing PDE operator is enforced. The stencil used here was called one stencil one face by Orsini et al. [20] and is shown in Figure 1 in the case of a two-dimensional triangular unstructured mesh.

Unlike the global Symmetric method, the approximated function is given by the equation (8), where $L_\xi$ and $B_\xi$ are the differential operators acting on $\Psi$ view as a function of $\vec{\xi}$, $r$ is the number of polynomial terms and $NT$ is the total number of points inside the stencil. By replacing the approximation function in the boundary condition (7) and the governing PDE (6) and evaluating the resultant formulae at the respective points, a linear system $A\alpha = B$ is obtained. The symmetric interpolation matrix $A$ is given by the equation (9) and the column vector $B$ by the expression (10).

$$\phi(\vec{x}) = \sum_{j=1}^{NP} \alpha_j \Psi(\|\vec{x} - \vec{\xi}_j\|) + \sum_{j=NP+1}^{NP+NB} \alpha_j B_\xi[\Psi(\|\vec{x} - \vec{\xi}_j\|)] + \sum_{j=NP+NB+1}^{NP+NB+NPDE} \alpha_j L_\xi[\Psi(\|\vec{x} - \vec{\xi}_j\|)] + \sum_{j=1}^{r} \alpha_{j+NT} P_{m-1}^{j}(\vec{x}) \quad (8)$$

$$A = \begin{pmatrix} \mathbf{\Psi} & B_\xi[\mathbf{\Psi}] & L_\xi[\mathbf{\Psi}] & P_{m-1} \\ B_x[\mathbf{\Psi}] & B_x B_\xi[\mathbf{\Psi}] & B_x L_\xi[\mathbf{\Psi}] & B_x[P_{m-1}] \\ L_x[\mathbf{\Psi}] & L_x B_\xi[\mathbf{\Psi}] & L_x L_\xi[\mathbf{\Psi}] & L_x[P_{m-1}] \\ P_{m-1}^T & B_x[P_{m-1}^T] & L_x[P_{m-1}^T] & 0 \end{pmatrix} \quad (9)$$

$$B = \begin{pmatrix} \phi_{vol} \\ g(\vec{x}) \\ f(\vec{x}) \\ 0 \end{pmatrix} \tag{10}$$

The above linear equation system can not be solved because the column vector $B$ includes, in $\phi_{vol}$, the $NP$ unknown values of the variable at the central nodes contained in the stencil. However, the coefficient $\alpha$ in terms of the unknown variable values can be used to interpolate the variable and its gradient at the locations required by the FVM discretization. To do this, the approximated gradient is obtained by deriving the function (8). Then, the approximated variable and its gradient can be expressed in matrix notation by equations (11) and (12). The matrices $C_1$ and $C_2$ are known from the geometrical information, since they include the RBFs and the resultant functions after applying operators to them.

$$\phi(\vec{x}) = [C_{1l}]^T [\alpha] \tag{11}$$

$$\frac{\partial \phi(\vec{x})}{\partial x_i} = [C_{2il}]^T [\alpha] \tag{12}$$

To solve the heat equation (1) by the CV-RBF method, it is necessary to replace the variable (11), in this case temperature, and its gradient (12) obtained by the local Hermitian Interpolation in the FVM discretized equation (4). The expression obtained from this procedure is the final CV-RBF discretized equation (13). By applying the expression (13) to each one of the finite volumes, a linear or non-linear system of equation is obtained.

$$\left( \sum_{l=1}^{m} \left(kC_{2i}^T n_i \Delta\partial\Omega\right)\Big|_{\vec{x}=\vec{x}_l} - \sum_{l=1}^{m} \left(\rho \hat{C}_p C_1^T v_i n_i \Delta\partial\Omega\right)\Big|_{\vec{x}=\vec{x}_l} \right) A^{-1} B - S(\vec{x}_p)\Delta\Omega = 0 \tag{13}$$

In the non-linear case, the Newton-Rapshon method is used to solve the resultant equation system by computing numerically the Jacobian matrix by the central difference scheme. To applied the Hermitian Interpolation, a linear operator $L$ has to be constructed from the governing equation. In the case of non-linear heat conduction, the Laplace operator is obtained according to the expression (14). The superindex $n$ refers to the quantity values at the present iteration.

$$L_x(T^n) = \frac{\partial^2 T^n}{\partial x_i \partial x_i} = \frac{1}{k(T^{n-1})} \left( -\frac{\partial k(T^{n-1})}{\partial T} \frac{\partial T^{n-1}}{\partial x_i} \frac{\partial T^{n-1}}{\partial x_i} - S(\vec{x}) \right) \tag{14}$$

## 4 Numerical results

The CV-RBF method is applied to solve the heat equation (1) on two dimensional domains for the linear and non-linear heat conduction in a solid, and to the thermal

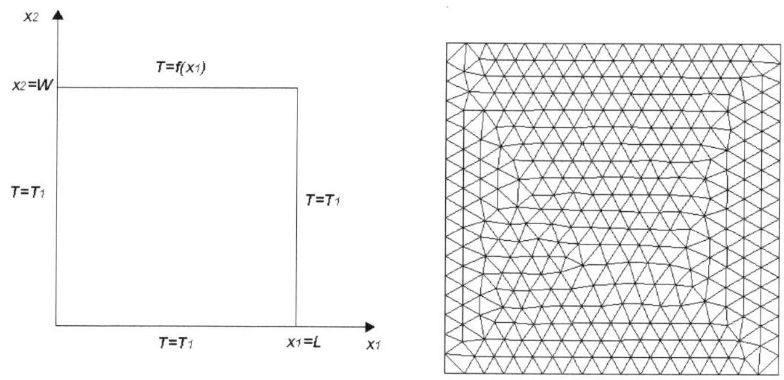


(a) (b)

Figure 2: Domain characteristics, (a) geometry, (b) unstructured mesh.

developing profile in the plane Poiseuille flow. The results are compared to the analytical solution for case of conduction, and to the numerical results found in literature for the convection problem.

### 4.1 Dirichlet problem

A rectangular domain of $W \times L$ which represents a solid thin plate is considered. In Figure 2 (a), the geometry and the boundary conditions are shown. The governing equation is reduced to the expression (15) in terms of the variable $\theta = T - T_1$ where $T_1$ is the temperature in all boundaries except at $x_2 = W$.

$$\frac{\partial}{\partial x_j}\left(k\frac{\partial \theta}{\partial x_j}\right) = -S(x_1, x_2) \tag{15}$$

The boundary condition at $x_2 = W$ is given by the expression (16) and the respective analytical solution is the equation (17), where $\theta_m = T_2 - T1$ and $T_2$ is the maximum temperature value. The shape parameter for the MQ function is equal to the mean distance between central and neighbor nodes.

$$f(x_1) = T_2 \sin\left(\frac{n\pi x_1}{L}\right) \tag{16}$$

$$\theta(x_1, x_2) = \theta_m \sin\left(\frac{\pi x_1}{L}\right)\frac{\sinh(\pi x_2/L)}{\sinh(\pi W/L)} \tag{17}$$

The unstructured mesh used on the domain, in the case of 384 volumes, is shown in Figure 2 (b). First, the numerical solutions, the global Root Mean Square (RMS) errors and the global maximum errors are obtained for different mesh densities. The results are presented in Table 1 where the convergence of the method can be seen for the linear conduction problem.

The numerical results obtained for a mesh of 384 volumes are shown in Figure 3, on three domain lines with $x_1$ constant. The relative error increases near

Table 1: RMS and maximum error in the solution of linear heat conduction problem.

| Mesh size | $\epsilon_{max}$ | $\epsilon_{RMS}$ |
|---|---|---|
| 32 vol | 0.0422 | 0.0165 |
| 100 vol | 0.0245 | 0.0105 |
| 384 vol | 0.0136 | 0.0042 |
| 882 vol | 0.0053 | 0.0013 |

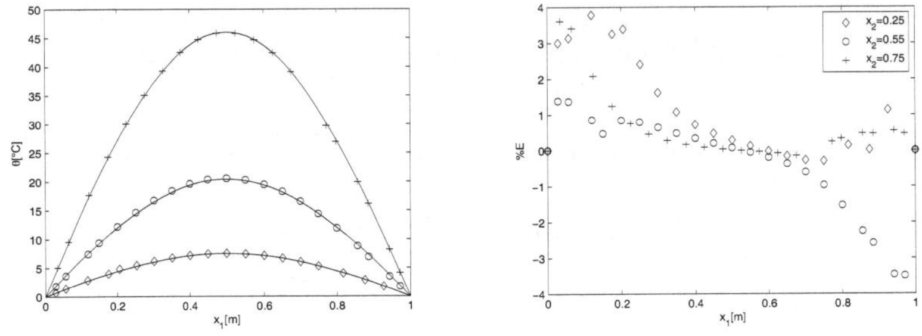

(a) (b)

Figure 3: Solution of the linear heat conduction problem, (a) temperature profile on lines $x_2$ =0.25 ($\diamond$), $x_2$ = 0.50 ($o$) and $x_2$ = 0.75 (+), (b) relative error.

boundaries but its magnitude is not greater than 4%. According to these results, the CV-RBF method can solve accurately the linear conduction problem with Dirichlet boundary conditions.

### 4.2 Non-linear conduction

A hypothetical heat conduction situation is considered in a solid material with a thermal conductivity in terms of temperature. The phenomena is modelled by the equation (15) where $\theta = T - T_1$, $k = T^{1.3}$ and $S$ is obtained by replacing the imposed solution (18) on the governing PDE. The problem geometry and the boundary conditions are the same as those presented in the Figure 2, with $f(x_1) = T_1$.

$$T(x_1, x_2) - T_\infty = 100\, T_\infty x_1 x_2 (1 - x_1)(1 - x_2) e^{-(x_1^2 + x_2^2)} \tag{18}$$

The numerical solution is obtained using the MQ function with $m = 1$ and a shape parameter $c = 0.1h$, where $h$ is the mean distance between central nodes in the stencil. A 30 × 30 structured mesh is used in the solution. The Figure 4 shows the exact solution contour (left) and the numerical solution contour (right).

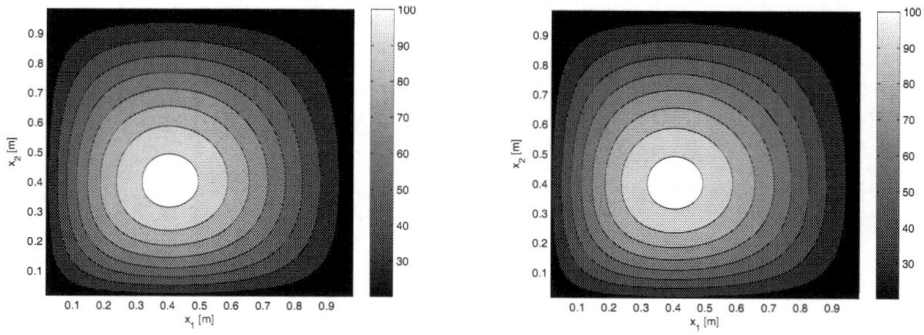


(a) (b)

Figure 4: Temperature contours for the non-linear heat conduction problem: (a) analytical solution; (b) numerical solution.

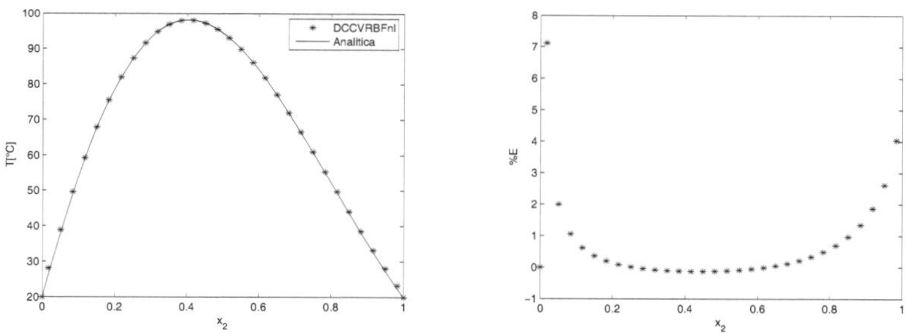


(a) (b)

Figure 5: Comparison between analytical and numerical solution of the non-linear heat conduction problem at the line $x_1 = 0.5$: (a) temperature profile; (b) relative error.

Also, the temperature profile and the relative error at the line $x_1 = 0.5$ are shown in Figures 5 (a) and (b), respectively. The figures mentioned show the good agreement between the analytical and numerical solution. The relative error, as in the linear case, increases near boundaries without exceeding $8\%$. This error is presented locally and does not affect the global RMS error which is $0.54\%$.

### 4.3 Temperature profile in the plane Poiseuille flow

An incompressible Newtonian fluid flows in the $x_1$ direction between two plates as shown in Figure 6. A hydraulic developed flow in steady state and a field velocity independent of temperature are assumed. According to this, the known velocity field is the plane Poiseuille flow described by equation (19). The fluid selected is air with constant properties and the maximum velocity is set to get a global Peclet number close to 1200.

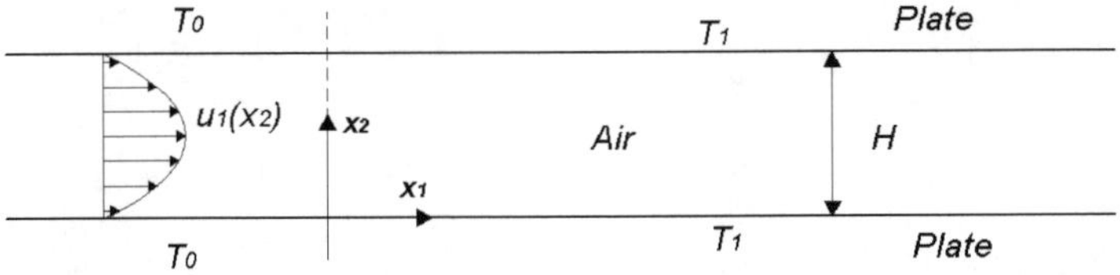


Figure 6: Geometrical configuration for the heat transfer in plane Poiseuille flow.

$$u_1(x_2) = (u_1)_{max}\left(1 - \left(\frac{x_2 - 0.5H}{0.5H}\right)^2\right) \tag{19}$$

The governing equation is the expression (1) without including the source term. The boundary conditions are given by the expressions from (20) to (22) and the domain is divided by a $30 \times 90$ structured mesh. The MQ with $m = 1$ is used and the shape parameter is computed as $c = 0.1h$. The analytical solution is known as the Graetz problem in rectangular coordinates, but there is not a close form of the solution. Therefore, the results are compared to those obtained by Divo and Kassab [22] employing an RBF collocation method.

$$T(0, x_2) = T_0 \tag{20}$$

$$T(x_1, 0) = T(x_1, H) = T_1 \tag{21}$$

$$\frac{\partial T(L, x_2)}{\partial x_1} = 0 \tag{22}$$

The thermal profile obtained is shown in Figure 7 (a). The temperature distribution agrees with the physical phenomena involved here, since near the plates the temperature is $T_1$ and it increases to $T_0$ close to the center line. In the Figure 7 (b) are presented the temperature profiles on several $x_1$ constant lines and can

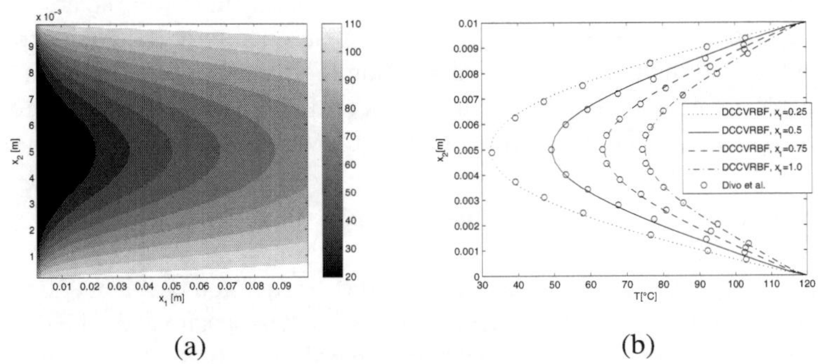


(a) (b)

Figure 7: Temperature profile in the plane Poiseuille flow: (a) Temperature contour; (b) Temperature profiles at $x_1$ constant lines.

be seen the agreement between the numerical results obtained and those found in literature.

## 5 Conclusions

The heat equation was solved accurately for basic two-dimensional situations by the CV-RBF method. This is a successful application of the method considering the equations solved and the coupling to the Newton-Rapshon method.

The CV-RBF method is an improvement to the traditional FVM due to the geometric treatment versatility. Both the results in unstructured and structured meshes are in agreement to the reference data.

The MQ function with $m = 1$ was used to solve accurately the heat equation. Although, the solution depends on the shape parameter and there is not all-purpose values or expressions for it. Hence, obtaining the shape parameter is an aspect to be improved in the method.

The main strengths of the CV-RBF method are the capability to deal with unstructured meshes, the high order of the interpolation scheme and the intrinsic upwind differencing.

## References


[1] Patankar, S., *Numerical heat transfer and fluid flow*. Computational and Physical Processes in Mechanics and Thermal Sciences, Taylor & Francis, 1980.

[2] Versteeg, H.K. & Malalasekera, W., *Introduction to computational fluid dynamics: the finite volume method*. Longman Scientific & Technical, 1995.

[3] Wesseling, P., *Principles of computational fluid dynamics*. Computational Mathematics, Springer, 2001.

[4] Abgrall, R., On essentially non-oscillatory schemes on unstructured meshes. *Journal of Computational Physics*, **114(1)**, pp. 45–98, 1994.

[5] Jayantha, P.A. & Turner, I.W., A second order control-volume-finite-element least-squares strategy for simulating diffusion in strongly anisotropic media. *Journal of Computational Physics*, **23(1)**, pp. 1–16, 2005.

[6] Truscott, S.L. & Turner, I.W., An investigation of the accuracy of the control volume-element based on triangular prismatic elements for simulating diffusion in anisotropic media, part b: Fundamentals. *Numerical Heat Transfer*, **46(1)**, pp. 243–268, 2004.

[7] Manzini, G. & Putti, M., Mesh locking effects in the finite volume solution of 2-d anisotropic difusin equations. *Journal of Computacional Physics*, **220(1)**, pp. 751–771, 2007.

[8] Wright, G. & Fornberg, B., Scattered node compact finite difference-type formulas generated from radial basis functions. *Journal of Computational Physics*, **212(1)**, 2006.

[9] Kansa, E.J., Multiquadrics -a scattered data approximation scheme with applications to computational fluid dynamics-ii solution to parabolic, hyperbolic and elliptic partial differential equations. *Computers & Mathematics with applications*, **19(8–9)**, pp. 127–145, 1990.

[10] Zerroukat, M., Power, H. & Chen, C.S., A numerical method for heat transfer problems using collocation and radial basis functions. *International Journal for Numerical Methods in Engineering*, **42(7)**, pp. 1263–1278, 1998.

[11] Li, J., Cheng, A.H.D. & Chen, C.S., A comparison of efficiency and error convergence of multiquadric collocation method and finite element method. *Engineering Analysis with Boundary Elements*, **27(3)**, pp. 251–257, 2003.

[12] Cheng, A.H.D. & Cabral, J.J.S.P., Direct solution of ill-posed boundary value problems by radial basis function collocation method. *International Journal for Numerical Methods in Engineering*, **64(1)**, pp. 45–64, 2005.

[13] Chantasiriwan, S., Cartesian grid methods using radial basis functions for solving poisson, helmholtz, and diffusionconvection equations. *Engineering Analysis with Boundary Elements*, **28(12)**, pp. 1417–1425, 2004.

[14] Kansa, E.J. & Hon, Y.C., Circumventing the ill- conditioning problem with multiquadric radial basis functions: Applications to elliptic partial differential equations. *Computers & Mathematics with applications*, **39(1)**, pp. 123–137, 2000.

[15] Fasshauer, G., Solving partial differential equations by collocation with radial basis functions. *Proceedings of Chamonix, Vanderbilt University Press*, pp. 1–8, 1997.

[16] Jumarhon, B., Amini, S. & Chen, K., The hermite collocation method using radial basis functions. *Engineering Analysis with Boundary Elements*, **24(7–8)**, pp. 607–611, 2000.

[17] Power, H. & Barraco, V., A comparison analysis between unsymmetric and symmetric radial basis function collocation methods for the numerical solution of partial differential equations. *Computers & Mathematics with Applications*, **43(3-5)**, pp. 551–583, 2002.

[18] Moroney, T.J. & Turner, I.W., A finite volume method based on radial basis functions for two-dimensional nonlinear diffusion equations. *Applied Mathematical Modelling*, **30(10)**, 2006.

[19] Moroney, T.J. & Turner, I.W., A three dimensional finite volume method based on radial basis functions for the accurate computacional modelling of nonlinear diffusion equations. *Journal of Computational Physics*, **225(2)**, 2007.

[20] Orsini, P., Power, H. & Morovan, H., Improving volume element methods by meshless radial basis function techniques. *Computer modelling in Engineering and Sciences*, **769(1)**, pp. 1–21, 2008.

[21] Fedoseyev, A.I., Friedman, M. & Kansa, E.J., Improve multiquadric method for elliptic partial differential equations via pde collocation on the boundary. *Computers & Mathematics with applications*, **43(3–5)**, pp. 439–455, 2002.

[22] Divo, E. & Kassab, A.J., A meshless method for conjugate heat problems. *Engineering Analysis with Boundary Elements*, **29(1)**, pp. 136–149, 2005.

# On the efficiency of the method of fundamental solutions for acoustic scattering by a poroelastic material

B. Nennig, E. Perrey-Debain & J.-D. Chazot
*Université Technologique de Compiègne, Laboratoire Roberval UMR CNRS 6253, France*

## Abstract

The Method of Fundamental Solutions is now a well established technique that has proved to be reliable for a specific range of wave problems such as the scattering of acoustic and elastic waves by obstacles and inclusions of regular shapes. The goal of this paper is to show that the technique can be extended in order to solve transmission problems whereby an incident acoustic pressure wave impinges on a poroelastic material of finite dimension. For homogeneous and isotropic materials, the wave equation for the fluid phase and solid phase displacements are found to be decoupled thanks to the Helmholtz decomposition. This allows a systematic way for obtaining an analytic expression for the fundamental solution describing the wave displacement field in the material. The efficiency of the technique relies on choosing an appropriate set of fundamental solutions as well as properly imposing the transmission conditions at the air-porous interface.

In this paper, we address this issue showing results involving bidimensional scatterers of various shapes. In particular, it is shown that reliable error indicators can be used to assess the quality of the results. Comparisons with results computed using a mixed pressure-displacement finite element formulation illustrate the great advantage of this new technique both in terms of computational resources and mesh preparation.
*Keywords: method of fundamental solutions, Biot's equations, poroelastic, porous material, scattering.*

WIT Transactions on Modelling and Simulation, Vol 50, © 2010 WIT Press
www.witpress.com, ISSN 1743-355X (on-line)
doi:10.2495/BE100161

# 1 Introduction

Poroelastic materials are often used for their good sound absorbing capabilities in the middle and high frequency range. Typical applications can be found in the context of the transport noise reduction or to enhance the quality of the room acoustics. The description of wave propagation in porous media is not limited to audible acoustics as the Biot's model [1] was originally developed for geologic applications. Because of their inherent diphasic features and the strong contrasts that may exist between the solid and the fluid phases, the waves propagation modeling remains a difficult task often leading to heavy computational costs. In the context of Finite Element Method (FEM), some developments have been proposed using Lagrange or hierarchical finite elements [2, 3, 4]. Because of the disparity of scale, the so-called poroelastic elements have a slower convergence than purely elastic or fluid element [3]. To make the matter worst, Biot's equations are frequency dependent and large FEM system matrices have to be recalculated for each frequency. For homogeneous and isotropic materials, the Boundary Element Method (BEM) offers an alternative [5]. The method has the great advantage to reduce the entire problem to only unknowns on the boundaries. However, the system matrix is full and there is still the need to discretize the boundary domain as well as performing regular and singular integrations over each boundary elements.

In the past decade, several researchers have focused their work on meshless methods in order to avoid the time-consuming problem originated in mesh generation for complicated geometries. In this regard, the Method of Fundamental Solution (MFS) has been shown to be efficient for solving a large variety of physical problems as long as a fundamental solution of the underlying differential equation(s) is known. In particular the MFS is suitable for scattering problems by choosing appropriate fundamental solutions satisfying the radiation condition at infinity. The method shares the same advantages as BEM over domain discretization methods because there is no need to create a mesh over the entire domain. Furthermore, as no integration is needed, some numerical difficulties encountered with BEM are avoided. For comprehensive reviews on applications of the MFS one can refer to Fairweather et *al.* [6, 7].

In this work, we are interested in applying as well as assessing the MFS for the simulation of the 2D acoustic wave scattering by a poroelastic material. After presenting the MFS formulation in Section 2, the method's performances are measured against an academic problem for which exact analytical solutions are available. In particular, it is shown that reliable error indicators can be used to assess the quality of the results. The last section shows practical applications with comparisons with a FE model.

# 2 Formulation of the method

Consider a time-harmonic acoustic plane wave $\varphi_0^{\mathrm{inc}} = \exp \mathrm{i}k_0\zeta \cdot \mathbf{x}$ (with the convention $\mathrm{e}^{-\mathrm{i}\omega t}$) in an unbounded exterior propagative domain $\Omega^e$ incident upon one (or more) poroelastic inclusion(s) denoted by $\Omega^i$ with boundary $\Gamma$ as shown in

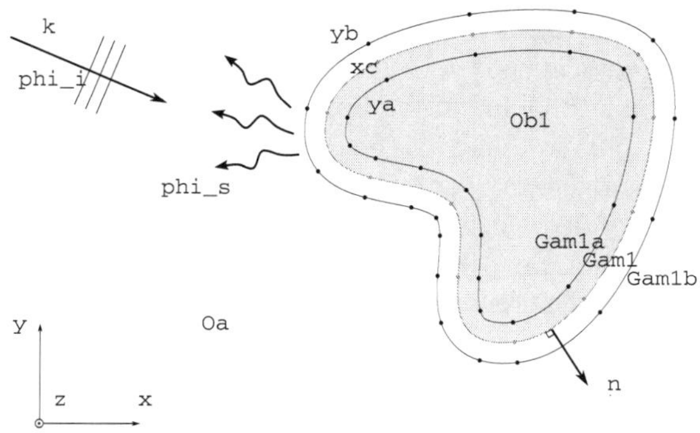


Figure 1: Scattering geometry.

Fig. 1. We call $\mathbf{n}$, the unit outward normal vector to $\Omega^i$. In the surrounding acoustic domain $\Omega^e$, the fluid is inviscid and the acoustic displacement potential $\varphi_0$ obeys the wave equation

$$\Delta\varphi_0 + k_0^2\varphi_0 = 0. \tag{1}$$

Here, $k_0 = \omega/c_0$ is the classical wavenumber defined as the ratio of the angular frequency $\omega$ and the sound speed $c_0$. To express the transmission conditions at the interface $\Gamma$, it is convenient to introduce the particle displacement perturbation $\mathbf{w} = \nabla\varphi_0$. With this definition, the acoustic pressure is obtained from the linearized momentum equation as $p = \rho_0\omega^2\varphi_0$. In (1) it is natural to split the potential into an incident and a scattered part as $\varphi_0 = \varphi_0^{\rm inc} + \varphi_0^{\rm sc}$ (and similarly for the pressure and displacement). Here we require the scattered field to satisfy the usual Sommerfeld radiation condition at infinity. In the poroelastic inclusions the acoustic waves propagation are described by Biot's model [1]. This latter is grounded on the superposition of a fluid phase and a solid phase which are coupled together and respectively described by the fluid phase displacement $\mathbf{U}$ and the solid phase displacement $\mathbf{u}$. For time-harmonic representation, we have the following coupled system [1]

$$\nabla\cdot\boldsymbol{\sigma}^s + \omega^2(\rho_{11}\mathbf{u} + \rho_{12}\mathbf{U}) = 0, \tag{2}$$

$$\nabla\cdot\boldsymbol{\sigma}^f + \omega^2(\rho_{12}\mathbf{u} + \rho_{22}\mathbf{U}) = 0. \tag{3}$$

Solid and fluid phase stress tensors are given by

$$\boldsymbol{\sigma}^s = (A\,\nabla\cdot\mathbf{u} + Q\,\nabla\cdot\mathbf{U})\,\mathbf{I} + 2N\,\boldsymbol{\varepsilon}^s, \tag{4a}$$

$$\boldsymbol{\sigma}^f = (Q\,\nabla\cdot\mathbf{u} + R\,\nabla\cdot\mathbf{U})\,\mathbf{I}, \tag{4b}$$

where $\boldsymbol{\varepsilon}^s = 1/2(\nabla\mathbf{u} + (\nabla\mathbf{u})^{\rm t})$ is the usual strain tensor and $\mathbf{I}$ is the identity matrix. The total stress tensor $\boldsymbol{\sigma}^t$ is, by definition, the sum of $\boldsymbol{\sigma}^f$ and $\boldsymbol{\sigma}^s$. Biot's coefficients $A, N, Q, R$ are related to the material properties by the Allard-Johnson model. Their expressions can be found in the literature or in the reference textbook [8] as well as the other quantities introduced in this section. These quantities are all complex and frequency-dependent, $A$ and $N$ correspond to the Lamé coefficients, $R$ is the effective bulk modulus of the fluid phase and $Q$ indicates the coupling of the two phases volumic dilatation. The imaginary part of $A$ and $N$ includes the structural damping and, in $Q$ and $R$ this part includes the thermal dissipation. The imaginary parts of the effective density coefficients $\rho_{11}$, $\rho_{22}$ and $\rho_{12}$ take into account viscous damping. Now, the complete solution to the problem is found after applying the classical air-porous transmission conditions [9, 10] on the interface $\Gamma$, i.e.

$$p_p - p^{\mathrm{sc}} = p^{\mathrm{inc}}\,, \tag{5a}$$

$$\phi \mathbf{U}\cdot\mathbf{n} + (1-\phi)\mathbf{u}\cdot\mathbf{n} - \mathbf{w}^{\mathrm{sc}}\cdot\mathbf{n} = \mathbf{w}^{\mathrm{inc}}\cdot\mathbf{n}\,, \tag{5b}$$

$$\boldsymbol{\sigma}^t\mathbf{n} + p^{\mathrm{sc}}\,\mathbf{n} = -p^{\mathrm{inc}}\,\mathbf{n}\,. \tag{5c}$$

Here $\phi$ is the porosity and the pore pressure $p_p$ is obtained from the fluid phase tensor as $-\mathbf{I}\phi p_p = \boldsymbol{\sigma}^f$.

For homogeneous and isotropic materials, the wave equation for the fluid phase and solid phase displacements are found to be decoupled thanks to the Helmholtz decomposition. Both solid and fluid displacement fields are then written as

$$\mathbf{u} = \nabla\varphi + \nabla\wedge(\varphi_3\cdot\mathbf{e}_3) \quad \text{and} \quad \mathbf{U} = \nabla\chi + \nabla\wedge(\Theta\cdot\mathbf{e}_3). \tag{6}$$

After equations decoupling, we have [8]: $\varphi = \varphi_1 + \varphi_2$ and $\chi = \mu_1\varphi_1 + \mu_2\varphi_2$ where

$$\mu_i = \frac{Pk_i^2 - \omega^2\rho_{11}}{\omega^2\rho_{12} - Qk_i^2}, \quad i = 1, 2, \tag{7}$$

are the waves amplitude ratios between the two phases in the porous material (here, $P = A + 2N$). Similarly, the potential $\Theta$ is simply obtained as $\Theta = \mu_3\varphi_3$ with $\mu_3 = \rho_{12}/\rho_{22}$. Under this form, each potential $\varphi_i$ ($i = 1, 2, 3$) fulfills the Helmholtz equation

$$\Delta\varphi_i + k_i^2\varphi_i = 0, \tag{8}$$

and the associated complex wavenumbers are given explicitly in [8]. Physically, there are two compressional waves associated with $\varphi_1$, $\varphi_2$ and one rotational (shear) wave associated with $\varphi_3$. They all propagate in the two phases and their relative contribution are given by the coefficients $\mu_i$. If such a decomposition holds in elastodynamics, the coexistence of two phases in the poroelastic media adds another fluid-born compressional wave which is not present is elastic solids.

The MFS theory starts by choosing an appropriate set of fundamental solutions for both propagative domains $\Omega^e$ and $\Omega^i$. In the acoustic domain, a natural choice is to pick these solutions using the well-known free field Green function, i.e. $G_0(\mathbf{x},\mathbf{y}) = \mathrm{i}/4H_0(k_0\,|\mathbf{x}-\mathbf{y}|)$. So we seek the scattered field as a distribution of monopole

$$p^{\mathrm{sc}}(\mathbf{x}) = \rho_0\omega^2 \sum_{q_0=1}^{Q_0} A^0_{q_0} G_0(\mathbf{x},\mathbf{y}^0_{q_0}), \tag{9}$$

where source points $\mathbf{y}^0_{q_0}$ are chosen to be located on $\hat{\Gamma}_0$ in $\Omega^i$. The displacement vector is obtained by simply taking the gradient. To simulate the wave field in the inclusion, a possible choice is to use fundamental solutions for poroelastic media which explicit form can be found, for instance, in [11]. An easier option is to construct these solutions by simply expanding each potential in the form

$$\varphi_i(\mathbf{x}) = \sum_{q_i=1}^{Q_i} A^i_{q_i} G_i(\mathbf{x},\mathbf{y}^i_{q_i}), \quad \text{for } i = 1,2,3, \tag{10}$$

where $G_i(\mathbf{x},\mathbf{y}) = \mathrm{i}/4H_0(k_i\,|\mathbf{x}-\mathbf{y}|)$ is the fundamental solution for the Helmholtz equation. Here, $Q_i$ is the number of source points $\mathbf{y}^i_{q_i}$ associated with the $i^{\mathrm{th}}$ potential and the $A^i_{q_i}$'s are unknown amplitudes. These points are located on a fictitious boundary $\hat{\Gamma}_i$ in the exterior domain $\Omega^e$ as shown in Fig. 1. The influence of the location of the source points on the quality of the solution will be discussed in the next section. Now, using these expansion in (6), we find the expression for the different physical quantities involved:

$$\phi p_p(\mathbf{x}) = \sum_{i=1}^{2} k_i^2(Q+\mu_i R) \sum_{q_i=1}^{Q_i} A^i_{q_i} G_i(\mathbf{x},\mathbf{y}^i_{q_i}) \tag{11a}$$

$$\mathbf{u}(\mathbf{x}) = \sum_{i=1}^{2}\sum_{q_i=1}^{Q_i} A^i_{q_i} \nabla G_i(\mathbf{x},\mathbf{y}^i_{q_i}) + \sum_{q_3=1}^{Q_3} A^3_{q_3} \nabla_\perp G_3(\mathbf{x},\mathbf{y}^3_{q_3}) \tag{11b}$$

$$\mathbf{U}(\mathbf{x}) = \sum_{i=1}^{2}\sum_{q_i=1}^{Q_i} A^i_{q_i} \mu_i \nabla G_i(\mathbf{x},\mathbf{y}^i_{q_i}) + \sum_{q_3=1}^{Q_3} A^3_{q_3} \mu_3 \nabla_\perp G_3(\mathbf{x},\mathbf{y}^3_{q_3}) \tag{11c}$$

where $\nabla_\perp \equiv (\partial_{x_2}, -\partial_{x_1})^{\mathrm{t}}$ stands for the orthogonal gradient operator. From these expressions and using (4), we finally get the normal total stress tensor $\boldsymbol{\sigma}^t$. The explicit forms for the tensor coefficients are too cumbersome to be inserted in this paper. Note that other type of solutions could have been considered (Green functions and/or their derivatives or plane waves for instance) and the present choice was mainly motivated for its simplicity. Furthermore, it has the great advantage of allowing us to specify independently the number of source points $Q_i$ for each kind of wave, this is not the case when using the Green function. In this respect, our method does not fall strictly into the category of the MPS since it does not rely on the Biot's equations displacements fundamental tensor.

Now, given a set of $N_{\rm col}$ collocation points $\mathbf{x}_l$ ($l = 1, \ldots, N_{\rm col}$). on $\Gamma$, substituting (11) in the transmission conditions (5) at each collocation point yields three linear system of the form

$$\mathbf{M}_\alpha \mathbf{A} = \mathbf{F}_\alpha^{\rm inc}, \quad \alpha = \rm a, b, c. \tag{12}$$

Here, subscript $\alpha$ refers to the type of condition involved (there is one system for each condition (5a), (5b), (5c)). The right hand side vector $\mathbf{F}_\alpha^{\rm inc}$ stems from the incident wave (pressure and displacement). The unknown vector $\mathbf{A}$ contains the amplitudes of all sources. Since the acoustic displacement is expected to behave like $|\mathbf{w}| \sim k_0\phi_0$, it is judicious to rescale the cinematic conditions (5b) by multiplying the associated lines by $\rho_0 c_0 \omega$. This yields the matrix system

$$\mathbf{M}\mathbf{A} = \mathbf{F}^{\rm inc}. \tag{13}$$

In this work, early results showed that it is preferable for reasons of stability to consider more collocation points than the number of unknowns $N_{\rm dof} = \sum_{i=0}^{3} Q_i$. To be more specific, it was found that taking $N_{\rm col} = 2N_{\rm dof}$ guarantees that results have converged, i.e. the numerical solution becomes insensitive to the number of collocation points. In the following, all calculations are performed using this ratio. Note that when dealing with 'nice' shaped inclusions (the circular scatterer for instance), the MFS has been observed to perform a bit better with interpolation schemes (i.e. $N_{\rm col} = N_{\rm dof}$) than least-square schemes ($N_{\rm col} > N_{\rm dof}$). However, it was found preferable for the sake of robustness and generality to favor the second approach.

After multiplying by the hermitian transpose $\mathbf{M}^\dagger$ we finally get

$$\mathbf{M}^\dagger\mathbf{M}\mathbf{A} = \mathbf{M}^\dagger\mathbf{F}^{\rm inc}. \tag{14}$$

The two advantages for doing this operation are (i) having to deal with a hermitian square system and (ii) that a simple error estimator is available. Indeed, coming back to the original problem (12), we can define the *a posteriori* error estimator for each conditions at the interface ($\alpha = \rm a, b, c$). So we put

$$\mathcal{E}_\alpha = 100\,\frac{\|\mathbf{M}_\alpha\mathbf{A} - \mathbf{F}_\alpha^{\rm inc}\|_2}{\|\mathbf{F}_\alpha^{\rm inc}\|_2}, \tag{15}$$

where $\mathbf{A}$ is solution of (14).

As for the choice for the source points positions, we follow Alves [12] and take these points along the discrete normal direction, so we put $\mathbf{y}_{q_i}^i = \hat{\mathbf{y}}_{q_i}^i + s^i\,\tilde{\mathbf{n}}_{q_i}^i$ where points $\hat{\mathbf{y}}_{q_i}^i$ ($q_i = 1, \ldots, Q_i$ and $i = 0, \ldots, 4$) are distributed on $\Gamma$ and $\tilde{\mathbf{n}}_{q_i}^i$ is the approximate normal vector defined by

$$\tilde{\mathbf{n}}_{q_i}^i = \frac{(\hat{\mathbf{y}}_{q_i+1}^i - \hat{\mathbf{y}}_{q_i-1}^i)^\perp}{\max_{q_i}|\hat{\mathbf{y}}_{q_i+1}^i - \hat{\mathbf{y}}_{q_i-1}^i|}. \tag{16}$$

Here, symbol $\perp$ signifies that we take the orthogonal vector pointing outward. The normalization is chosen here for pure convenience to make sure that the normal vectors amplitudes never exceed the unity and that coefficients $s^i$ correspond to the farthest distance from the interface $\Gamma$.

## 3 Validations for the circular-shaped inclusion

We shall validate and assess the efficiency of the method on the specific case of the scattering of a horizontal acoustic plane wave by a circular-shaped poroelastic inclusion. In polar co-ordinates, the inner and outer wave fields can be represented by separable solutions. Each potential is given by infinite series which are well behaved allowing us to produce very accurate results without deterioration at high frequency. The derivation of the exact form for these solutions is quite lengthy and this will not be inserted in the present paper. For a proper assessment of the method's performance, it is convenient to define the relative error on the boundary for each physical quantity (call it $X$):

$$\hat{\mathcal{E}}(X) = 100 \frac{\left\| \mathbf{X} - \hat{\mathbf{X}} \right\|_2}{\left\| \hat{\mathbf{X}} \right\|_2}, \tag{17}$$

where $X$ represents either the acoustic pressure, the fluid phase or solid phase normal displacement or the normal stress at the interface and $\mathbf{X}$ is the associated vector containing the value of $X$ at the collocation points. Here, $\hat{\mathbf{X}}$ stands for a reference solution vector, either obtained analytically in the present case or numerically computed with another method when analytical solutions are not available. Now, the aim of this section is to identify the effects of the main parameters of the problem, namely the sources locations and the number of sources. For this latter, we are still left with the problem of finding a quasi-optimal relationship between the $Q_i$'s. Through extremely intensive calculations, not shown here, we found that choosing the same number for each wave type was probably the best option offering the best trade-off between accuracy, conditioning and simplicity. This greatly simplifies the analysis as we can now put $Q = Q_i$ for $i = 0, \ldots, 4$ and perform the analysis with a single parameter $Q$. Similarly, we take the points $\hat{\mathbf{y}}^i_{q_i}$ all equal and put $s^i = s$ with $s^0 = -s$. Note this choice seems to be against results given in [13] where it is advocated that the best option, when using plane waves, is precisely not to take the same number of wave directions for each type of waves (that is shear and pressure wave types in the elasticity case) and this is particularly relevant when the ratio between wavenumbers is large. The reasons for this probably lie in that, plane waves and singular sources behave differently in terms of their approximation properties and it would be interesting to explore this further via numerical experiments in the spirit of [14] for instance.

For the sake of illustration we chose a polymer foam (XFM) commonly used in the transport industry. The materials properties are reported in Tab. 1. To give an idea of the complex wavenumbers for the wave potentials in the material, we find that $k_1 \approx 17.6 + 1.21\mathrm{i}$, $k_2 \approx 21.4 + 14.1\mathrm{i}$ and $k_3 \approx 38.9 + 2.1\mathrm{i}$ (these values corresponds to the frequency 500 Hz). The corresponding acoustic wavenumber is

Table 1: Materials properties for the XFM foam. All quantities reported are defined in [9].

| $\phi$ | $\sigma$ | $\alpha_{\text{inf}}$ | $\Lambda$ | $\Lambda'$ | $\rho_1$ | $N$ | $\nu$ |
|---|---|---|---|---|---|---|---|
| 0.98 | 13.5 | 1.7 | 80 | 160 | 30 | 200(1 - 0.05i) | 0.35 |

$k_0 \approx 9.2$. On Fig. 2 (a) are shown the influence of the source locations as well as the number of sources (i.e. $Q$) on the normal solid displacement error $u_n = \mathbf{u} \cdot \mathbf{n}$, i.e. $\hat{\mathcal{E}}(u_n)$ and the *a posteriori* error $\mathcal{E}_{\text{b}}$. The evolution of the conditioning (in the 2-norm) of the associated system is also shown (Fig. 2 (b)). Before we comment on these results, we should point out that we chose to measure the normal displacement error rather than the pressure error for the simple reason that it was observed, that a small error on the displacement normally guarantees a smaller error on the pressure (though the opposite is not true). The convergence curves together with the conditioning curves typically illustrate the MFS paradigm which many authors have already observed and discussed: ill conditioned systems lead to accurate results. In the present situation, it is concluded that the further away from the boundary the source points are (i.e. the higher $s$), the better the results as long as the conditioning of the system does not exceed a certain value above which results are likely to be corrupted by round off errors. This value, say $\text{cond}_2 \approx 10^{16}$, is in line with standard double precision arithmetic. Note that, in our algorithm, matrices are inverted using the `pinv` function from MATLAB so that for very ill-conditioned matrices ($\text{cond}_2 > 10^{16}$) the SVD solver is automatically used with some thresholding in order to dampen the effects of the round-off errors. In this

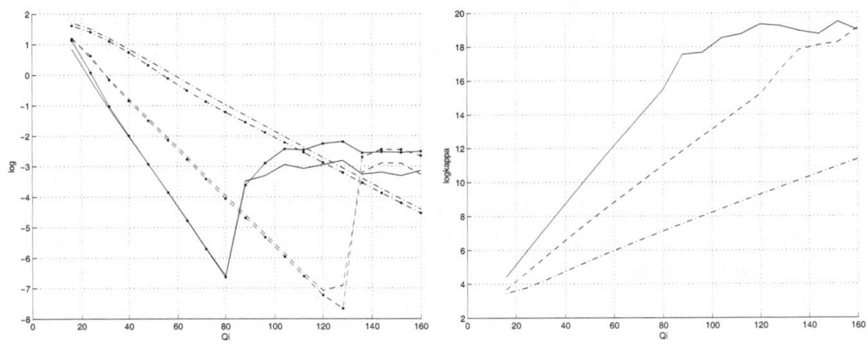


Figure 2: Effects of the sources location on the error and the conditioning (–·– $s = 0.1a$, - - $s = 0.2a$, —— $s = 0.3a$). (a): error on the solid phase normal displacement for the XFM foam at 1500Hz; line with markers · refers to $\hat{\mathcal{E}}(u_n)$ error and the simple line to the *a posteriori* error $\mathcal{E}_{\text{b}}$. (b): conditioning (2-norm) of $\mathbf{M}^{\dagger}\mathbf{M}$.

matter, we think that better results could perhaps be made by properly filtering the small singular values. Another approach would be to apply the Tikhonov regularization in order to speed up the computation.

In fact, the most striking feature of these convergence curves is the almost perfect adequation between the true error $\hat{\mathcal{E}}(u_n)$ and the its algebraic counterpart, the *a posteriori* error $\mathcal{E}_b$ and through lots of numerical experiments, we always observe very good correlations. The mathematical reasons for this goes beyond the scope of this paper but we believe that in most cases the algebraic *a posteriori* errors $\mathcal{E}_\alpha$ should serve as good error indicators especially when reference solutions are not available.

## 4 Numerical examples and concluding remarks

This section describes two examples illustrating the efficiency of the MFS. The first shows the MFS performances for inclusions presenting geometrical singularities such as corners and in the second example, we tackle multiple scattering problems with non-convex shaped objects. Because there are no analytical solutions to these problems, all reference solutions are computed using Finite Element models. These computations are carried using Lagrange quadratic finite element in both fluid and poroelastic domains. The $(\mathbf{u}, p_p)$ formulation of Biot's equations [2, 9] is used and the non reflecting boundary conditions are implemented using the Bermudez's perfectly matched layer formulation [15] (PML). As for the accuracy of these FE reference solutions, numerical tests carried out on the circular shaped scattering problem shows that these results are reliable up to around 1 percent of error on the solid phase normal displacement which is acceptable if engineering accuracy is sought. The first example concerns that of a square-shaped poroelastic inclusion of side length $a = 0.2$ m. The incident plane wave is horizontal travelling in the $\zeta = (1, 0)$ direction and the frequency is 1500 Hz. In this scenario, the boundary curve is not analytic and source points around the square are placed along the normal to the boundary avoiding the corners, see Fig. 3 (a).

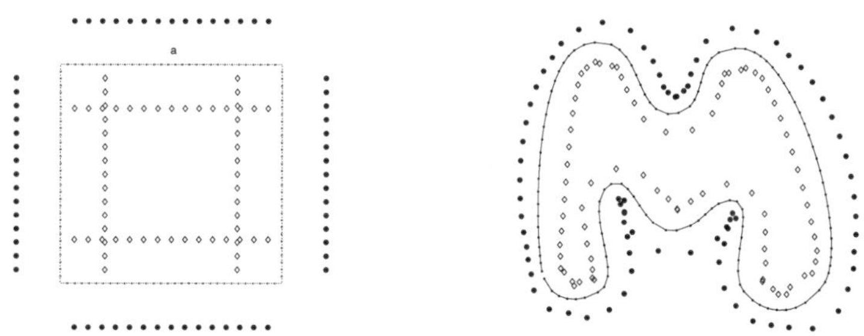


Figure 3: (a) Square-shaped inclusion. (b) The 'M' shaped scatterer. Here symbols $+$, $\times$, $\circ$ indicate the location of the source points of $\Omega^e$ and $\diamond$ those of $\Omega^i$. The collocations point on $\Gamma$ are identified by the $\cdot$ markers.

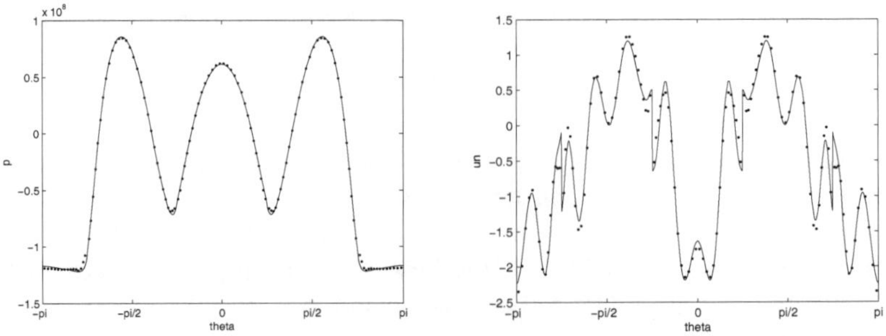


Figure 4: (a): Real part of the pressure field on Γ at 1500 Hz for the square-shaped inclusion; solid line: FEM solution and dots refers to the MFS solution. (b): Real part of solid phase normal displacement.

In this respect, we found that putting more sources in the vicinity of the corners did not show any improvement. Comparisons between the MFS and the FE results are conveniently shown in Fig. 4. Here, the abscissae $\theta$ refers the usual polar coordinate of the collocation points on the boundary (the square is centered at the origin). As expected, the discrepancies are maximum near the corners and this is particularly noticeable for the displacement curve. The associated global errors are in good agreements as we found $\mathcal{E}_{\rm b} = 4.42\%$ with $\hat{\mathcal{E}}(u_n) = 1.88\%$ at 500 Hz and $\mathcal{E}_{\rm b} = 4.95\%$ with $\hat{\mathcal{E}}(u_n) = 4.31\%$ at 1500 Hz. Here we took $Q = 60$ ($N_{\rm dof} = 240$).

The last numerical experiment concerns a multiple scattering problem with non-convex objects having the shape of the three letters 'M', 'F' and 'S'. The boundaries have been drawn using a graphical software and interpolated with Bezier curves. Fig. 3 (b) shows the location of the source points as well as the collocation points on the 'M' letter. The example considered here is that of an incident plane wave travelling in the $\zeta = (\cos \pi/4, -\sin \pi/4)$ direction and the frequency is 6000 Hz. In our calculation, we took $Q = 80$ for each scatterer; this yields $N_{\rm dof} = 3 * 4 * Q = 960$ variables. The *a posteriori* errors for the pressure and the displacement are $\mathcal{E}_{\rm a} = 1.78\%$ and $\mathcal{E}_{\rm b} = 2.77\%$. The total pressure in the acoustic domain and in the porous inclusions is shown in Fig. 5. The absorbing properties of the porous material can be visually identified. Note that the CPU time needed for computing the matrix coefficients and inverting the system does not exceed 20 s on a single PC (Intel Core 2 Duo T7100 @ 1.80 GHz) and the use of the FEM would be far more demanding in order to achieve the same level of accuracy.

Through these examples, we showed the great advantage of the MFS both in terms of computational resources and mesh preparation. When dealing with scatterers presenting geometrical singularities such as corners, comparisons with FE computations showed that few percent errors results can still be achieved which is acceptable if engineering accuracy is sought. It is believed that this could have a positive impact for solving a wide range of noise control problems ranging from

Figure 5: Real part of the total pressure in the acoustic domain and in the porous inclusions.

multiple scattering to shape foam optimization. Work is ongoing by the authors to extend the method to 3D configurations.

## References

[1] Biot, M.A., Theory of propagation of elastic waves in a fluid-saturated porous solid. I. low-frequency range. II. higher frequency range. *J Acoust Soc Am*, **28(2)**, pp. 168–191, 1956.

[2] Atalla, N., Hamdi, M.A. & Panneton, R., Enhanced weak integral formulation for the mixed $(u, p)$ poroelastic equations. *J Acoust Soc Am*, **109(6)**, pp. 3065–3068, 2001.

[3] Rigobert, S., Atalla, N. & Sgard, F., Investigation of the convergence of the mixed displacement-pressure formulation for three-dimensional poroelastic materials using hierarchical elements. *J Acoust Soc Am*, **114(5)**, pp. 2607–2617, 2003.

[4] Horlin, N.E., Nordstron, M. & Goransson, P., A 3-D hierarchical fe formulation of biot's equations for elasto-acoustic modeling of porous media. *J Sound Vib*, **245(4)**, pp. 633–652, 2001.

[5] Tanneau, O., Lamary, P. & Chevalier, Y., A boundary element method for porous media. *J Acoust Soc Am*, **20(3)**, pp. 1239–1251, 2006.

[6] Fairweather, G., Karageorghis, A. & Martin, P.A., The method of fundamental solutions for scattering and radiation problems. *Engng Anal Bound Elem*, **27**, pp. 759–769, 2003.

[7] Fairweather, G. & Karageorghis, A., The method of fundamental solutions for elliptic boundary value problems. *Adv Comput Math*, **9(1-2)**, pp. 69–95, 1998.

[8] Allard, J.F., *Propagation of Sound in Porous Media: Modeling Sound Absorbing Materials*. Chapman & Hall, 1993.

[9] Debergue, P., Panneton, R. & Atalla, N., Boundary conditions for the weak formulation of the mixed $(u, p)$ poroelasticity problem. *J Acoust Soc Am*, **106(5)**, pp. 2393–2390, 1999.

[10] Lovera, O.L., Boundary conditions for a fluid-saturated porous solid. *Geophysics*, **52(2)**, pp. 174–178, 1987.

[11] Bonnet, G., Basic singular solutions for a poroelastic medium in the dynamic range. *J Acoust Soc Am*, **82(5)**, pp. 1758–1762, 1987.

[12] Alves, C.J.S., On the choice of source points in the method of fundamental solutions. *Engng Anal Bound Elem*, **33**, pp. 1348–1361, 2009.

[13] Perrey-Debain, E., Plane wave decomposition in the unit disc: Convergence estimates and computational aspects. *J Comput Appl Math*, **193(1)**, pp. 140–156, 2006.

[14] Alves, C.J.S. & Valtchev, S.S., Numerical comparison of two meshfree methods for acoustic wave scattering. *Engng Anal Bound Elem*, **29**, pp. 371–382, 2005.

[15] Bermudez, A., Hervella-Nieto, L., Prieto, A. & Rodriguez, R., An optimal perfectly matched layer with unbounded absorbing function for time-harmonic acoustic scattering problems. *J Comput Phy*, **223**, pp. 469–488, 2007.

# The method of fundamental solutions for inverse obstacle acoustic scattering

A. Karageorghis[1] & D. Lesnic[2]
[1]*Department of Mathematics and Statistics, University of Nicosia, Cyprus*
[2]*Department of Applied Mathematics, University of Leeds, UK*

## Abstract

In this paper we propose a simple method for detecting (shape, size and location) a scatterer embedded in a host acoustic homogeneous medium from scant measurements of the scattered acoustic pressure in the vicinity (near-or far-field) of the obstacle. We develop a nonlinear constrained minimization regularized method of fundamental solutions for obtaining the numerical solution of the inverse problem. The stability of the numerical scheme is investigated by inverting measurements contaminated with noise.
*Keywords: inverse acoustic scattering, method of fundamental solutions.*

## 1 Introduction

The inverse problems of time-harmonic acoustic scattering of waves from obstacles of arbitrary shape embedded in a fluid, or solid medium have been of considerable interest to researchers for many years, see e.g. [1]. In addition to being of academic interest, these problems have physical applications in the fields of radar and sonar detection; the ability to "see" in real time in complete darkness in murky water for deep sea submarines, underwater surveillance and target acquisition, detection of objects in the ocean, either fully submerged or partially buried in the seafloor, ultrasound medical imaging of soft tissues, nondestructive testing of materials, etc.

The inverse problem we consider in this paper is to determine the boundary of an insonified scatterer from scant measurements of its response when excited by impinging plane waves. The scatterer can be sound-soft, sound-hard or convectively embedded in the full-space. Although some uniqueness results are known, notably in [2], the problem is still difficult to solve since it is nonlinear,

WIT Transactions on Modelling and Simulation, Vol 50, © 2010 WIT Press
www.witpress.com, ISSN 1743-355X (on-line)
doi:10.2495/BE100171

ill-posed (unstable), and computationally intensive in the high-frequency regime, see [3].

Integral equation methods, see e.g. [4], provide a basic tool in scattering theory mainly due to the fact that the formulation of obstacle scattering problems leads to boundary value problems defined over unbounded domains. Hence, their formulation in terms of boundary integral equations not only reduces by one the dimensionality of the problem, but also replaces a problem over an unbounded domain by one over a bounded contour, or surface. From a numerical point of view both these advantages have made integral equation methods, such as the boundary element method (BEM) one of the most powerful technique for the approximate solution of obstacle scattering in homogeneous media, see [5].

The method of fundamental solutions (MFS) may be viewed as a meshless BEM in which the governing field is approximated by a linear combination of fundamental solutions for the Helmholtz equation whose singularities are placed outside the solution domain. In contrast to the BEM, the MFS provides accurate approximations directly at points arbitrarily close to the boundary. Moreover, since no shape functions or elements are involved, it is easier to implement than the BEM. Also, the fictitious eigenfrequency difficulty with occurs with the BEM for the exterior Helmholtz problem apparently is not present in the MFS, [6]. Up to now, the MFS has never been applied to inverse obstacle scattering, although its principal idea of representing the solution as a single-layer potential has been used previously in [7].

In this paper, first, the direct, linear and well-posed problem of acoustic scattering for time-harmonic waves from impenetrable obstacles is approached using the MFS to find the scattered wave. Then, using a random noisy perturbation of the scattered wave as input data, the inverse, nonlinear and ill-posed problem of determining the boundary of the obstacle is approached based on a regularized optimization procedure which uses the MFS solver at each iteration until a prescribed stopping criterion is satisfied. A similar approach using the BEM instead of MFS has been previously investigated in [8], but only for sound-hard obstacles parameterised by a maximum of seven parameters and for exact scattered wave data only.

The method proposed in this paper can be easily extended to penetrable obstacles, as well as to electromagnetic scattering problems.

## 2 The direct problem

We are concerned with the scattering of a time harmonic acoustic wave by a bounded impenetrable obstacle $D \subset \mathbb{R}^n$. The wave motion can then be determined from the potential $u$ which satisfies the Helmholtz equation

$$\Delta u + k^2 u = 0 \quad \text{in} \quad \mathbb{R}^n \backslash \bar{D}, \tag{1}$$

where $k > 0$ is the wave number.

For the direct (forward) scattering problem, we are given the wave number $k > 0$, a bounded region $D \subset \mathbb{R}^n$ with boundary $\partial D \in C^2$ such that $\mathbb{R}^n \backslash \bar{D}$ is connected, and an incident field given by a plane wave moving in the unit direction $\widehat{\boldsymbol{d}}$, namely

$$u^{\text{inc}}(\boldsymbol{x}) = \mathrm{e}^{ik\boldsymbol{x}\cdot\widehat{\boldsymbol{d}}}. \tag{2}$$

Then the scattering of $u^{\text{inc}}$ by $D$ produces a scattered wave $u^{\text{s}}$ with corresponding total field $u = u^{\text{s}} + u^{\text{inc}}$. The direct scattering problem is to determine the scattered pressure $u^{\text{s}} \in C^2(\mathbb{R}^n\backslash\overline{D}) \cap C^1(\mathbb{R}^n\backslash D)$ which satisfies the Helmholtz equation (1), i.e.

$$\Delta u^{\text{s}}(\boldsymbol{x}) + k^2 u^{\text{s}}(\boldsymbol{x}) = 0, \qquad \boldsymbol{x} \in \mathbb{R}^n\backslash\overline{D}, \tag{3}$$

the Sommerfeld infinity radiation condition

$$\lim_{R\to\infty} R^{(n-1)/2}\Big(\frac{\partial u^{\text{s}}}{\partial R}(\boldsymbol{x}) - iku^{\text{s}}(\boldsymbol{x})\Big) = 0 \tag{4}$$

uniformly with respect to $\widehat{\boldsymbol{x}} = \boldsymbol{x}/|\boldsymbol{x}|$, where $R = |\boldsymbol{x}|$, and the boundary conditions corresponding to:

(a) a sound-soft obstacle (Dirichlet boundary condition)

$$u(\boldsymbol{x}) = 0 \quad \Longleftrightarrow \quad u^{\text{s}}(\boldsymbol{x}) = -\mathrm{e}^{ik\boldsymbol{x}\cdot\widehat{\boldsymbol{d}}}, \quad \boldsymbol{x} \in \partial D. \tag{5}$$

(b) a sound-hard obstacle (Neumann boundary condition)

$$\frac{\partial u}{\partial \nu}(\boldsymbol{x}) = 0 \quad \Longleftrightarrow \quad \frac{\partial u^{\text{s}}}{\partial \nu}(\boldsymbol{x}) = -\frac{\partial(\mathrm{e}^{ik\boldsymbol{x}\cdot\widehat{\boldsymbol{d}}})}{\partial \nu}(\boldsymbol{x}), \quad \boldsymbol{x} \in \partial D, \tag{6}$$

where $\boldsymbol{\nu}$ is the outward unit normal to $\partial D$.

(c) a convective obstacle (impedance, Robin boundary condition)

$$\frac{\partial u}{\partial \nu}(\boldsymbol{x}) + i\lambda u(\boldsymbol{x}) = 0, \quad \Longleftrightarrow \quad \frac{\partial u^{\text{s}}}{\partial \nu}(\boldsymbol{x}) + i\lambda u^{\text{s}}(\boldsymbol{x})$$
$$= -\frac{\partial(\mathrm{e}^{ik\boldsymbol{x}\cdot\widehat{\boldsymbol{d}}})}{\partial \nu}(\boldsymbol{x}) - i\lambda \mathrm{e}^{ik\boldsymbol{x}\cdot\widehat{\boldsymbol{d}}}, \quad \boldsymbol{x} \in \partial D, \tag{7}$$

where $\lambda \in \mathbb{C}$, $\Re(\lambda) \geq 0$ is the impedance parameter.

In the above model we have assumed that the obstacle $D$ is impenetrable. The case of penetrable obstacles which leads to a transmission problem will be investigated elsewhere.

The unique solvability of the direct problems (3)-(5), (3), (4) and (6), (3), (4) and (7) is well-established, see e.g. [1]. Furthermore, one has the following Green's representation formula

$$\eta(\boldsymbol{x})u^{\text{s}}(\boldsymbol{x}) = \int_{\partial D} \left[ u^{\text{s}}(\boldsymbol{y}) \frac{\partial G_n}{\partial \nu(\boldsymbol{y})}(\boldsymbol{x}, \boldsymbol{y}) - G_n(\boldsymbol{x}, \boldsymbol{y}) \frac{\partial u^{\text{s}}}{\partial \nu}(\boldsymbol{y}) \right] ds(\boldsymbol{y}), \quad \boldsymbol{x} \in \mathbb{R}^n\backslash D,$$

where $\eta(\boldsymbol{x}) = 1$ if $\boldsymbol{x} \in \mathbb{R}^n \backslash \overline{D}$ and $\eta(\boldsymbol{x}) = \frac{1}{2}$ if $\boldsymbol{x} \in \partial D$, and $G_n$ is a fundamental solution of the Helmholtz equation defined by

$$G_n(\boldsymbol{x}, \boldsymbol{y}) = \begin{cases} \dfrac{i}{4} H_0^{(1)}(k|\boldsymbol{x} - \boldsymbol{y}|), & n = 2, \\ \\ \dfrac{\mathrm{e}^{ik|\boldsymbol{x}-\boldsymbol{y}|}}{4\pi|\boldsymbol{x} - \boldsymbol{y}|}, & n = 3, \end{cases} \tag{8}$$

where $H_0^{(1)}$ denotes the Hänkel function of the first kind and of order zero.

## 3 The far-field pattern

The radiation condition (4) leads to an asymptotic behaviour for the scattered wave of the form

$$u^{\mathrm{s}}(\boldsymbol{x}) = \frac{\mathrm{e}^{ikR}}{R^{(n-1)/2}} \{u_\infty(\widehat{\boldsymbol{x}}) + O(R^{-1})\}, \quad R \to \infty$$

uniformly with respect to $\widehat{\boldsymbol{x}}$, where the function $u_\infty$ is called far-field pattern of the scattered wave. The following representations for the far-field pattern are given in [9].

(a) In the case of the Dirichlet boundary condition (5) we have

$$u_\infty(\widehat{\boldsymbol{x}}) = -\gamma_n \int_{\partial D} \frac{\partial u}{\partial \nu}(\boldsymbol{y}) \mathrm{e}^{-ik\widehat{\boldsymbol{x}}\cdot\boldsymbol{y}} ds(\boldsymbol{y}). \tag{9}$$

(b) In the case of the Neumann boundary condition (6) we have

$$u_\infty(\widehat{\boldsymbol{x}}) = \gamma_n \int_{\partial D} u(\boldsymbol{y}) \frac{\partial}{\partial \nu}(\mathrm{e}^{-ik\widehat{\boldsymbol{x}}\cdot\boldsymbol{y}}) ds(\boldsymbol{y}). \tag{10}$$

(c) In the case of the Robin boundary condition (7) we have

$$u_\infty(\widehat{\boldsymbol{x}}) = \gamma_n \int_{\partial D} u(y) \left[ \frac{\partial}{\partial \nu}(\mathrm{e}^{-ik\widehat{\boldsymbol{x}}\cdot\boldsymbol{y}}) + i\lambda \mathrm{e}^{-ik\widehat{\boldsymbol{x}}\cdot\boldsymbol{y}} \right] ds(\boldsymbol{y}), \tag{11}$$

where $\gamma_2 = (1+i)/(4\sqrt{k\pi})$ and $\gamma_3 = 1/(4\pi)$.

## 4 Inverse problems

In this section we recall several uniqueness results for inverse obstacle scattering problems.

**Inverse Problem 1 (IP1)**

Determine $D$ from the knowledge of $u_\infty(\widehat{\boldsymbol{x}}, \widehat{\boldsymbol{d}})$ for all $\widehat{\boldsymbol{x}}, \widehat{\boldsymbol{d}} \in S^{n-1}$ and for one fixed wave number $k_0 > 0$.

**Theorem 1.** ([10]). Let $u^i$ be the total fields corresponding to sound-soft or -hard obstacles $D_i$ for $i = 1, 2$. If $u^1_\infty(\widehat{\boldsymbol{x}}; \widehat{\boldsymbol{d}}) = u^2_\infty(\widehat{\boldsymbol{x}}; \widehat{\boldsymbol{d}})$ for all $\widehat{\boldsymbol{x}}, \widehat{\boldsymbol{d}} \in S^{n-1}$, then $D_1 = D_2$.

 
www.witpress.com, ISSN 1743-355X (on-line)

**Inverse Problem 2 (IP2)**
Determine $D$ from the knowledge of $u_\infty(\widehat{\boldsymbol{x}}; \widehat{\boldsymbol{d}_0})$ for all $\widehat{\boldsymbol{x}} \in S^{n-1}$, for one fixed wave number $k_0 > 0$ and for one fixed direction $\widehat{\boldsymbol{d}_0} \in S^{n-1}$.
**Theorem 2.** ([2]). Let $u^i$ be the total fields corresponding to sound-soft obstacles $D_i \subset B(0, \zeta_n)$ for $i = 1, 2$, where $\zeta_2 = 2.40482/k$ and $\zeta_3 = \pi/k$. If $u^1_\infty(\widehat{\boldsymbol{x}}) = u^2_\infty(\widehat{\boldsymbol{x}})$ for all $\widehat{\boldsymbol{x}} \in S^{n-1}$, then $D_1 = D_2$.

**Inverse Problem 3 (IP3)**
Determine $D$ from the knowledge of $u^{\mathrm{s}}$ on some closed surface $\Gamma$ containing $D$ in its interior.

By the uniqueness for the exterior Dirichlet and Neumann problems, knowing $u^{\mathrm{s}}$ an a closed surface $\Gamma$ containing $D$ implies knowing the far-field pattern $u_\infty$ of $u^{\mathrm{s}}$. Therefore, the uniqueness results of Theorems 1 and 2 for the reconstruction from far-field data immediately carry over to the case of near-field data.

Although Theorems 1 and 2 ensure the uniqueness of the solution of the IP1, IP2 and IP3, the inverse problems are still ill-posed (unstable), as can be seen from the following theorem.
**Theorem 3.** ([7]). For a fixed incident field $u^{\mathrm{inc}}$, the mapping $\partial D \mapsto u_\infty$ is continuous from $C^1$ into $L^2(S^{n-1})$.

Therefore, one has to solve the nonlinear and ill-posed operator equation

$$F(\partial D) = u_\infty \tag{12}$$

for the unknown boundary $\partial D \in C^1$ by standard inversion methods, [5].

The IP1 has been treated in detail in [11] and therefore, in this study only the IP2 and IP3 will be investigated.

## 5 The method of fundamental solutions (MFS)

Based on the density results of Bogomolny [12], in the MFS we seek the solution of Helmholtz equation (3) in the form, see e.g. [13],

$$u^{\mathrm{s}}(\boldsymbol{x}) = \sum_{j=1}^{N} c_j G_n(\boldsymbol{x}, \boldsymbol{y}_j), \qquad \boldsymbol{x} \in \mathbb{R}^n \backslash \overline{D}, \tag{13}$$

where $\boldsymbol{y}_j \in D$ are singularities located in $D$, which are assumed fixed and prescribed, and $c_j$ are unknown coefficients to be determined by imposing the boundary conditions (5), (6) or (7).

We consider the inverse problem IP3 for simplicity, and choose as measurements

$$u^{\mathrm{s}}(\boldsymbol{x}_m) = \phi_m, \quad \frac{\partial u^{\mathrm{s}}}{\partial \nu}(\boldsymbol{x}_m) = \psi_m, \qquad m = \overline{1, M}, \tag{14}$$

where $(\boldsymbol{x}_m)_{m=\overline{1,M}}$ are uniformly distributed on the surface of the ball $B(\mathbf{0}, \zeta_n)$. Assuming that the obstacle $D$ is star-shaped, we also consider $N$ distributed points

$\tilde{\boldsymbol{x}}_m$ on $\partial D$ characterized by the radii $r_m = r(\widehat{\boldsymbol{x}}_m)$ for $m = \overline{1,N}$ on which we apply the boundary conditions

$$u^{\mathrm{s}}(\tilde{\boldsymbol{x}}_m) = -\mathrm{e}^{ik\tilde{\boldsymbol{x}}_m\cdot\widehat{\boldsymbol{d}}}, \tag{15}$$

or

$$\frac{\partial u^{\mathrm{s}}}{\partial \nu}(\tilde{\boldsymbol{x}}_m) = -\frac{\partial}{\partial \nu}\mathrm{e}^{ik\tilde{\boldsymbol{x}}_m\cdot\widehat{\boldsymbol{d}}}, \tag{16}$$

or

$$\frac{\partial u^{\mathrm{s}}}{\partial \nu}(\tilde{\boldsymbol{x}}_m) + i\lambda u^{\mathrm{s}}(\tilde{\boldsymbol{x}}_m) = -\frac{\partial}{\partial \nu}(\mathrm{e}^{ik\tilde{\boldsymbol{x}}_m\cdot\widehat{\boldsymbol{d}}}) - i\lambda \mathrm{e}^{ik\tilde{\boldsymbol{x}}_m\cdot\widehat{\boldsymbol{d}}}. \tag{17}$$

We also choose the singularities located at the $N$ uniformly distributed points $\boldsymbol{y}_j$ in $D$ characterized by radii $r(\boldsymbol{y}_j) = \eta r_j$, $j = \overline{1,N}$, where $\eta \in (0,1)$ is an appropriately chosen constant. Since $D \subset B(\mathbf{0}, \zeta_n)$ this also ensures that $\boldsymbol{y}_j$ are not situated on a closed curve $\Gamma_0$ for which $-k^2$ could be an eigenvalue for the Dirichlet problem for the Laplace equation in the interior domain enclosed by $\Gamma_0$. Therefore, we have $2N$ unknowns $(r_m)_{m=\overline{1,N}}$ and $(c_j)_{j=\overline{1,N}}$.

Imposing the sound-soft boundary condition (15) say, results in $N$ equations, namely,

$$-\mathrm{e}^{ik\tilde{\boldsymbol{x}}_m\cdot\hat{\boldsymbol{d}}} = \sum_{j=1}^{N} c_j G_n(\tilde{\boldsymbol{x}}_m, \boldsymbol{y}_j), \qquad m = \overline{1,N}. \tag{18}$$

Imposing the measurement conditions (14) results in $2M$ equations, namely,

$$\phi_m = \sum_{j=1}^{N} c_j G_n(\boldsymbol{x}_m, \boldsymbol{y}_j), \quad \psi_m = \sum_{j=1}^{N} c_j \frac{\partial G_n}{\partial \nu}(\boldsymbol{x}_m, \boldsymbol{y}_j), \quad m = \overline{1,M}. \tag{19}$$

### 5.1 Two-dimensional implementation

The two-dimensional MFS implementation is similar to that developed by the authors in [14] for the electrical impedance tomography in electrostatics. Let $\tilde{\theta}_m = 2\pi m/N$ for $m = \overline{0,N}$, be a uniform discretisation of the interval $[0, 2\pi]$. Assuming that the obstacle $D$ is star-shaped with respect to the origin, we consider the unknown boundary $\partial D$ given in parametric form by the points

$$\tilde{\boldsymbol{x}}_m = (r_m \cos(\tilde{\theta}_m), r_m \sin(\tilde{\theta}_m)), \qquad m = \overline{1,N}, \tag{20}$$

where $r_m > 0$ for $m = \overline{1,N}$ are unknown. The measurements (14) are given on the circle $B(\mathbf{0}, \zeta_2)$ which has the parametric form

$$\boldsymbol{x}_m = (\zeta_2 \cos(\theta_m), \zeta_2 \sin(\theta_m)), \qquad m = \overline{1,M}, \tag{21}$$

where $\theta_m = 2\pi m/M$ for $m = \overline{0,M}$. The singularities in $D$ are given by

$$\boldsymbol{y}_j = (\eta\, r_j \cos(\theta_j), \eta\, r_j \sin(\theta_j)), \qquad j = \overline{1,N}. \tag{22}$$

Then we minimize the penalized least-squares functional

$$T_{\mu}(\boldsymbol{c},\boldsymbol{r}) := \sum_{m=1}^{N}\left|\sum_{j=1}^{N} c_j G_2(\tilde{\boldsymbol{x}}_m,\boldsymbol{y}_j) + \mathrm{e}^{ik\tilde{\boldsymbol{x}}_m\cdot\hat{\boldsymbol{d}}}\right|^2$$

$$+\sum_{m=1}^{M}\left|\sum_{j=1}^{N} c_j G_2(\boldsymbol{x}_m,\boldsymbol{y}_j) - \phi_m\right|^2 + \sum_{m=1}^{M}\left|\sum_{j=1}^{N} c_j \frac{\partial G_2}{\partial \nu}(\boldsymbol{x}_m,\boldsymbol{y}_j) - \psi_m\right|^2$$

$$+\mu_1\sum_{j=1}^{N}|c_j|^2 + \mu_2\sum_{\ell=2}^{M}\left(r_\ell - r_{\ell-1}\right)^2, \tag{23}$$

where $\mu_1, \mu_2 > 0$ are regularization parameters, subject to the simple bounds on the variables

$$0 < r_m < \zeta_2, \qquad m = \overline{1,N}. \tag{24}$$

This optimization problem is accomplished using the `MINPACK` [15] routine `lmdif` which minimizes the unconstrained sum of squares of nonlinear functions and which does not require supplying the gradient of the functional (23). The constraints (24) are imposed during the iterative procedure by adjustment at each iteration. The minimization process terminates when either a user-specified tolerance is achieved or when a user-specified maximum number of function evaluations, `maxfev`, is reached. Thus in some cases, it is possible that the actual number of function evaluations preformed, `nfev`, is less than `maxfev`. In all numerical experiments carried out we set the tolerance to be equal to $10^{-10}$. The initial guess is only constrained to (24) and we have taken it to be a circle of radius $r_0$ centered at the origin.

The calculation of the fundamental solution $G_2$ in (8) and its derivatives is carried out using the code `MJY01A` from [16].

In three-dimensions one can use spherical coordinates.

# 6 Numerical results and discussion

We consider the simple case of a plane wave of wave number $k = 1$ forming an angle of $\alpha = 0$ with the $x$-axis impinging on a circular sound-soft scatterer of unit radius $B(\mathbf{0}; 1)$. The measurements (14) are taken on the circle of radius $\zeta_2 = 2.40428$. The exact solution for the direct scattering from a sound-soft circular obstacle $B(\mathbf{0}; a)$ satisfying equations (1), (2), (4) and (5) is given by, see e.g., [1],

$$u(r,\theta) = J_0(kr) - \frac{J_0(ka)H_0^{(1)}(kr)}{H_0^{(1)}(ka)} +$$
$$2\sum_{\ell=1}^{\infty} i^\ell \cos(\ell\theta)\left[J_\ell(kr) - \frac{J_\ell(ka)H_\ell^{(1)}(kr)}{H_\ell^{(1)}(ka)}\right],$$
$$(r,\theta) \in \mathbb{R}^2 \backslash B(\mathbf{0};a),\ r \geq 1,\ \theta \in [0, 2\pi), \tag{25}$$

where $H_\ell^{(1)}$ and $J_\ell$ are the Hänkel and Bessel functions of the first kind of order $\ell$, respectively. Note that the expression (25) is calculated by means of the routine MJYNA from [16]. We take $r_0 = 0.6$ and $\eta = 0.3$. In order to investigate the stability of the numerical solution, we fix $M = N = 64$ and vary the amount of noise $p$ included in the input data (14).

In Figure 1 we present the plots of the reconstructed boundary for noise $p = 5\%$, no regularization and with various numbers of `nfev`. It can be observed that as `nfev` increases so does the instability. In Figure 2 we present the plots of the reconstructed boundary for various values of the regularization parameter $\mu_1$ in (23) for $\mu_2 = 0$ and $p = 5\%$. In all cases `maxfev` was set to $5 \times 10^5$. Interestingly, the effect of increasing the value of $\mu_1$ was to reduce the number of `nfev` where convergence was reached. Improved stable results were obtained for $\mu_1 = O(10^{-4}) \div O(10^{-2})$. In Figure 3 we present the plots of the reconstructed boundary for various values of the regularization parameter $\mu_2$ in (23) for $\mu_1 = 0$, $p = 5\%$ and `maxfev`$= 5 \times 10^5$. Improved stable results were obtained for $\mu_2 = O(10^{-2}) \div O(1)$.

## 7 Conclusions

In this paper, it has been shown that the MFS is well-suited for the solution of inverse obstacle problems arising in acoustic scattering. The numerical experiments yield accurate results for exact data, but instabilities appear when

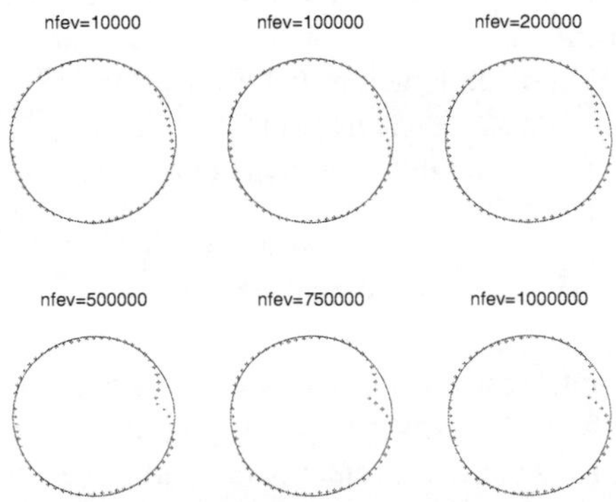


Figure 1: Results for various values of `nfev` for $p = 5\%$ and no regularization.

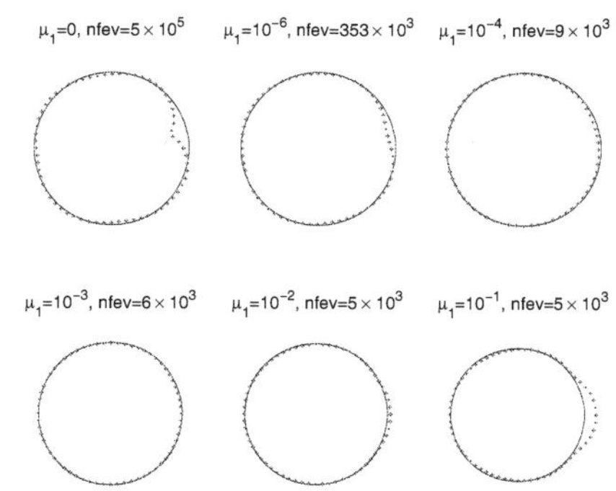


Figure 2: Results for various values of $\mu_1$ for $p = 5\%$, when $\mu_2 = 0$.

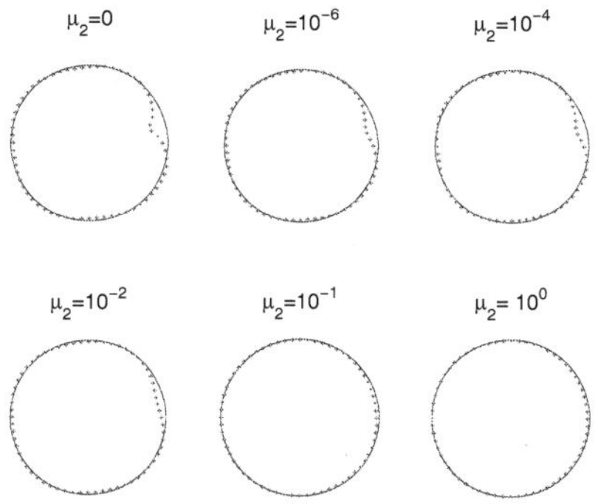


Figure 3: Results for various values of $\mu_2$ for $p = 5\%$, when $\mu_1 = 0$ and nfev= $5 \times 10^5$.

noise is introduced into the input data. Regularization can be achieved either by appropriately limiting the number of function evaluations or by introducing penalty terms in the objective cost functional which is minimized. The retrieval of more complicated shapes, multiple scatterers and the implementation of a corresponding three-dimensional MFS algorithm will be the subject of future work.

## Acknowledgement

The authors would like to thank the University of Cyprus for supporting this research.

## References


[1] Colton, D. & Kress, R., *Inverse acoustic and electromagnetic scattering theory*, volume 93 of *Applied Mathematical Sciences*. Springer-Verlag: Berlin, 2nd edition, 1998.

[2] Colton, D. & Sleeman, B.D., Uniqueness theorems for the inverse problem of acoustic scattering. *IMA J Appl Math*, **31(3)**, pp. 253–259, 1983.

[3] Kress, R., Integral equation methods in inverse acoustic and electromagnetic scattering. *Boundary integral formulations for inverse analysis*, Comput. Mech.: Southampton, pp. 67–92, 1997.

[4] Johansson, T. & Sleeman, B.D., Reconstruction of an acoustically sound-soft obstacle from one incident field and the far-field pattern. *IMA J Appl Math*, **72(1)**, pp. 96–112, 2007.

[5] Kress, R., Integral equation method in inverse obstacle scattering. *Eng Anal Boundary Elem*, **15**, pp. 171–179, 1995.

[6] Kondapalli, P., Shippy, D.J. & G. Fairweather, G., Analysis of acoustic scattering in fluids solids by the method of fundamental solutions. *J Acoust Soc Am*, **91**, pp. 1844–1854, 1992.

[7] Kress, R. & Zinn, A., On the numerical solution of the three-dimensional inverse obstacle scattering problem. *J Comput Appl Math*, **42(1)**, pp. 49–61, 1992.

[8] Na, S.W. & Kallivokas, L.F., Shape detection in inverse acoustic scattering problems. *16th ASCE Engineering Mechanics Conference*, University of Washington: Seattle, July 16-18, 2003.

[9] Kirsch, A., *An introduction to the mathematical theory of inverse problems*, volume 120 of *Applied Mathematical Sciences*. Springer-Verlag: New York, 1996.

[10] Kirsch, A. & Kress, R., Uniqueness in inverse obstacle scattering. *Inverse Problems*, **9(2)**, pp. 285–299, 1993.

[11] Colton, D., Coyle, J. & Monk, P., Recent developments in inverse acoustic scattering theory. *SIAM Rev*, **42(3)**, pp. 369–414, 2000.

[12] Bogomolny, A., Fundamental solutions method for elliptic boundary value problems. *SIAM J Numer Anal*, **22(4)**, pp. 644–669, 1985.

[13] Fairweather, G., Karageorghis, A. & Martin, P.A., The method of fundamental solutions for scattering and radiation problems. *Eng Anal Boundary Elem*, **27**, pp. 759–769, 2003.

[14] Karageorghis, A. & Lesnic, D., Detection of cavities using the method of fundamental solutions. *Inverse Problems Sci Eng*, **17(6)**, pp. 803–820, 2009.

[15] Garbow, B.S., E., H.K. & Moré, J.J., *MINPACK Project*. Argonne National Laboratory, 1980.

[16] Zhang, S. & Jin, J., *Computation of special functions*. A Wiley-Interscience Publication, John Wiley & Sons Inc.: New York, 1996.


 
www.witpress.com, ISSN 1743-355X (on-line)

# Free and forced vibrations of the shells of revolution interacting with the liquid

E. Strelnikova, E. Yeseleva, V. Gnitko & V. Naumenko
*Institute for Mechanical Engineering Problems of the Ukrainian Academy of Sciences, Ukraine*

## Abstract

The coupled problem of free and forced vibration of shells of revolution interacting with the liquid is under consideration. The problem was reduced to singular integral equations. The main objective of this paper is to present a detailed study of free and forced vibrations of a shell of revolution with an arbitrary meridian. For solving the boundary value problem of determining the fluid pressure on the shell, the boundary element method (BEM) is used. This method substantially reduces the computer time for the analysis and reveals new qualitative possibilities in modeling the dynamic behavior of shells. Numerical investigations of natural frequencies and mode shapes of the cylindrical tank with the incompressible fluid have been carried out. The behavior of structures subjected to a dynamic loading and interacting with the fluid is investigated.
*Keywords: free vibrations, forced vibrations, fluid-shell interaction, boundary and finite element methods.*

## 1 Introduction

Fluid-structure interaction has been studied extensively in the last decade [1–4]. The urgency of the problem is connected with wide application of the liquid storage tanks in engineering. The liquid storage tanks are important components of lifeline and industrial facilities. They are critical elements in municipal water supply and fire fighting systems, and in many industrial facilities for storage of water, oil, chemicals and liquefied natural gas, etc.

Several complexities are involved in the strength analysis of such tanks. If the excitation frequency is near the natural one, the high dynamic pressures may be produced on the tank walls due to the resonance. So calculation of the natural

WIT Transactions on Modelling and Simulation, Vol 50, © 2010 WIT Press
www.witpress.com, ISSN 1743-355X (on-line)
doi:10.2495/BE100181

frequencies and mode shapes is of great importance. Containers partially filled with the liquid are often subjected to an intensive external excitation. Problems associated with the dynamic behavior of the liquid storage tanks involve the analysis of two systems: the tank and the liquid, as well as the interaction between them along their boundaries. Due to the complexity of the problem, most of the original studies were experimental in nature. Several simplified theoretical investigations were also conducted and a few of these studies have been used as a basis for current design standards.

In [5, 6] authors offer the approach based on using the boundary element method (BEM) to the problem of natural vibrations of the fluid-filled elastic shells of revolution filled by the liquid, as well as to the problem of natural liquid vibrations in the rigid vessels. This approach has the certain advantages. In the basic equations the functions and their derivatives will be defined on the domain boundaries only. That allows reducing per unit dimension of the problem. In the present work the BEM technique is developed to research the processes of fluid-structure interaction of the shells of revolution at the dynamic loading. In the present paper the new effective method is elaborated. This method gives new qualitative possibilities in modeling the dynamic coupled problems.

## 2 Problem statement

Let us consider the problem of free and forced vibrations of an elastic shell of revolution with an arbitrary meridian and partially filled with the ideal incompressible fluid. We shall indicate a moistened surface of a shell through $S_1$. Let refer the Cartesian coordinate system $0xyz$ connected with a shell. The free surface of the liquid $S_0$ coincides with the plane $x0y$ in unperturbed state.

Suppose that the fluid is ideal and incompressible and the flow induced by vibrations of the shell is irrotational and consider small shell and fluid vibrations. Let $\boldsymbol{U}$ is the vector-function of the shell displacements and $\boldsymbol{V}$ is the fluid velocity vector.

Under these suppositions, there exists a velocity potential φ defined as

$$V_x = \frac{\partial \varphi}{\partial x}, \quad V_y = \frac{\partial \varphi}{\partial y}, \quad V_z = \frac{\partial \varphi}{\partial z}.$$

Assuming that the natural velocity of the fluid is zero, the pressure value, according to the Cauchy-Lagrange integral, can be represented as follows

$$P_l = -\rho_l \frac{\partial \varphi}{\partial t} . \tag{1}$$

Here $P_l$ is hydrodynamic pressure in the disturbed motion, and $\rho_l$ is the fluid density. A system of governing equations of motion of elastic shell with the liquid in the operator form is given by

$$\boldsymbol{LU} + \boldsymbol{M}\ddot{\boldsymbol{U}} = P_l + Q , \tag{2}$$

where $\boldsymbol{L}$, $\boldsymbol{M}$ are operators of elastic and mass forces of the shell; $\boldsymbol{U} = (u_1, u_2, w)$ is the displacement vector; $Q(t)$ is the vector of external surface load.

The kinematical boundary condition of the continuous fluid motion on the wetted shell surface $S_1$ can be represented as follows:

$$\boldsymbol{Vn} = \dot{\boldsymbol{U}}\boldsymbol{n} = \dot{w}\,, \tag{3}$$

where $\boldsymbol{n}$ is an external unit normal to wetted surface of the shell and $w$ is the shell normal displacement.

Let $z = H$ is the coordinate of non-disturbed free surface $S_0$ of the fluid. The boundary condition on $S_0$ corresponds to the requirement of pressure absence on the free surface. It can be written as follows:

$$P_l = 0\,. \tag{4}$$

The hydrodynamic problem (1)–(4) can be described by using the only unknown function φ $(x, y, z, t)$. To determine the velocity potential we obtain the following boundary value problem:

$$\boldsymbol{LU} + \boldsymbol{M}\ddot{\boldsymbol{U}} + \rho\dot{\varphi} = Q\,, \tag{5}$$

$$\nabla^2\varphi = 0\,, \tag{6}$$

$$\frac{\partial\varphi}{\partial n} = \dot{w}\,,\ \ P \in S_1\,, \tag{7}$$

$$\dot{\varphi} = 0\,,\ \ P \in S_0\,. \tag{8}$$

Here it is supposed, that $S = S_1 \cup S_0$ and point $P$ belongs to surface $S$.

## 3 System of the boundary integral equations

To solve the coupled hydro-elasticity problem it is necessary to determine the fluid pressure on the wetted surface of the shell. This problem is here reduced to the solution of the system of singular integral equations. In this section it is supposed, that velocity of the elastic shell is given. Determination of the fluid pressure on the wetted surface of the shell leads to the mixed boundary value problem for the Laplace equations (5)–(8).

We will seek the natural modes of vibration in the fluid in the form

$$U(x, y, z, t) = \sum_{k=1}^{m} u_k(x, y, z)\, c_k(t), \tag{9}$$

$$\varphi(x, y, z, t) = \sum_{k=1}^{m} \varphi_k(x, y, z)\, \dot{c}_k(t). \tag{10}$$

Here functions $u_k(x, y, z)$ are modes of natural vibrations in vacuum, $c_k(t)$ are unknown factors. To determine the set of functions $\varphi_k$ the following boundary value problems are formulated:

$$\nabla^2\varphi_k = 0\,,\ \ \frac{\partial\varphi_k}{\partial n} = w_k\,,\ \ P_0 \in S_1\,, \tag{11}$$

$$\varphi_k = 0\,,\ \ P_0 \in S_0\,, \qquad k = 1, m\,.$$

We will seek the harmonic functions $\varphi_k$ in the form of the sum of simple and double layer potentials [7]. The direct formulation of boundary integral equation method is in use. Dropping the index $k$, we have

$$2\pi\varphi(P_0) = \iint_S \frac{\partial\varphi}{\partial n}\frac{1}{|P-P_0|}dS - \iint_S \varphi\frac{\partial}{\partial n}\frac{1}{|P-P_0|}dS . \tag{12}$$

For the mixed problem described by eqns (5)–(8) this integral representation leads to the following system of singular integral equations:

$$2\pi\varphi(P_0) + \iint_{S_1} \varphi\frac{\partial}{\partial n}\frac{1}{|P-P_0|}dS_1 - \iint_{S_0} q\frac{1}{|P-P_0|}dS_0 =$$

$$= \iint_{S_1} w\frac{1}{|P-P_0|}dS_1, \quad P_0 \in S_1 , \tag{13}$$

$$\iint_{S_1} \varphi\frac{\partial}{\partial n}\frac{1}{|P-P_0|}dS_1 - \iint_{S_0} q\frac{1}{|P-P_0|}dS_0 = \iint_{S_1} w\frac{1}{|P-P_0|}dS_1, \quad P_0 \in S_0 ,$$

with regard to unknown functions $\varphi$ and $q$. Here function $\varphi$ defined on the surface $S_1$, represents the pressure on the wetted surface of the shell, and function $q$, defined on the surface $S_0$, is the normal component of the fluid velocity on the free surface.

We use furthermore the cylindrical coordinate system and represent unknown functions as Fourier series by circumferential coordinate

$$w = w(r,z)\cos\alpha\theta , \quad \varphi = \varphi(r,z)\cos\alpha\theta . \tag{14}$$

We will provide transformation in the same way as in [7]. The purpose is to bring the kernels in eqn (13) to standard elliptic integrals. Since

$$|P-P_0| = \sqrt{r^2 + r_0^2 + (z-z_0)^2 - 2rr_0\cos(\theta-\theta_0)} ,$$

$$a = r^2 + r_0^2 + (z-z_0)^2 , \quad b = 2rr_0 ,$$

we obtain the following expression

$$\frac{\partial}{\partial n}\frac{1}{|P-P_0|} - \frac{(\overrightarrow{P-P_0})\cdot\vec{n}}{|P-P_0|^3} = -\frac{n_r[r - r_0\cos(\theta-\theta_0)] + n_z(z-z_0)}{\left(\sqrt{a - b\cos(\theta-\theta_0)}\right)^3} .$$

As the result of these transformations we have obtained the system of integral equations with one-dimensional integrals computed only along the meridian of the wetted surface of the shell and radius of the fluid free surface.

This system takes the following form

$$2\pi\varphi(z_0) + \int_\Gamma \varphi(z)Q(z,z_0)r(z)d\Gamma - \int_0^R q(\rho)\Psi(P,P_0)\rho d\rho =$$

$$= \int_{\Gamma} w(z)\Psi(P,P_0)r(z)d\Gamma_1, \quad P_0 \in S_1,$$

$$\int_{\Gamma} \varphi(z)Q(z,z_0)r(z)d\Gamma - \int_{0}^{R} q(\rho)\Psi(P,P_0)\rho d\rho = \int_{\Gamma} w(z)\Psi(P,P_0)r(z)d\Gamma_1, \quad P_0 \in S_0.$$

Here

$$Q(z,z_0) = \frac{4}{\sqrt{a+b}}\left\{\frac{1}{2r}\left[\frac{r^2 - r_0^2 + (z_0 - z)^2}{a-b}\mathsf{E}_\alpha(k) - \mathsf{F}_\alpha(k)\right]n_r + \frac{z_0 - z}{a-b}\mathsf{E}_\alpha(k)n_z\right\},$$

$$\Psi(P,P_0) = \frac{4}{\sqrt{a+b}}\mathsf{F}_\alpha(k).$$

Here the following notations are introduced

$$\mathsf{E}_\alpha(k) = (-1)^\alpha\left(1 - 4\alpha^2\right)\int_0^{\pi/2} \cos 2\alpha\psi\sqrt{1 - k^2\sin^2\psi}\,d\psi,$$

$$\mathsf{F}_\alpha(k) = (-1)^\alpha \int_0^{\pi/2} \frac{\cos 2\alpha\psi d\psi}{\sqrt{1 - k^2\sin^2\psi}}, \qquad k^2 = \frac{2b}{a+b}.$$

Letting $\alpha = 0$ in the above expression, we obtain the standard elliptic first and second kind integrals.

For numerical solution of system described by eqns (13) the boundary element method with constant approximation of unknown density on elements was used.

# 4 The mode superposition method for coupled dynamic problems

After definition of functions $\varphi_k$ we substitute expressions (9), (10) in eqn (5) and obtain the following equation

$$\boldsymbol{L}\left(\sum_{k=1}^{m} c_k u_k\right) + \boldsymbol{M}\left(\sum_{k=1}^{m} \ddot{c}_k u_k\right) = -\rho\sum_{k=1}^{m} \ddot{c}_k\varphi_k + Q. \tag{15}$$

Let us $\omega_k$, $u_k$ are natural frequencies and free vibrations mode shapes of the shell in vacuum. For these the following relationships are valid

$$\boldsymbol{L}u_k = \omega_k^2 \boldsymbol{M}u_k, \quad (\boldsymbol{M}u_k, u_j) = \delta_{kj}. \tag{16}$$

Considering the result of dot product of eqn (15) by $u_j$ and taking into account relationship (16), we come to the set of second order differential equations about unknown factors $c_k$

$$\omega_j^2\delta_{kj}c_j + \delta_{kj}\ddot{c}_j = -\rho\sum_{k=1}^{m}\ddot{c}_k\varphi_k u_j + Qu_j, \quad j = 1, m. \tag{17}$$

To solve this system the Runge-Kutta method was used. Natural frequencies and modes of the shell vibrations in vacuum were defined by the finite element method.

## 5 Numerical results

Let us consider a cylindrical shell with a flat bottom partially filled with the fluid. The geometry of the tank is shown in Figure 1 and the parameters are following: the radius is $R = 1$ m, the thickness is $h = 0.01$ m, the length $L = 2$ m, Young's modulus $E = 2\cdot 10^5$ MPa, Poisson's ratio $\nu = 0.3$, the material's density is $\rho = 7800$ kg/m$^3$, the fluid density $\rho_l = 1000$ kg/m$^3$. The filling level of the fluid is denoted as $H$. Boundary conditions are following: $u_r = u_z = u_\theta = 0$ to $z = 0$ and $r = R$.

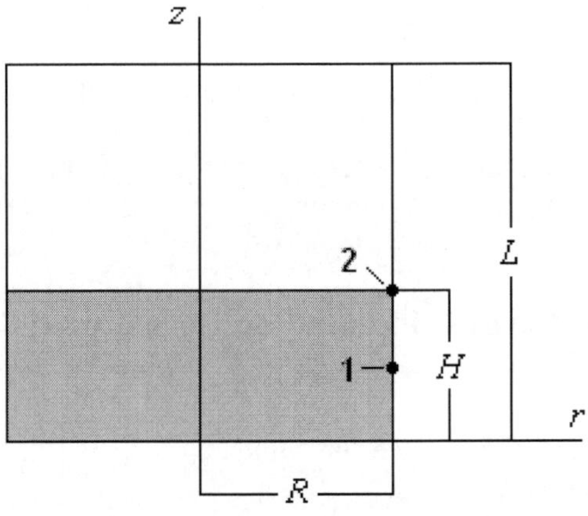


Figure 1: Cylindrical tank with the fluid.

At first we analyze free vibrations of the empty shell and the shell with the liquid. In the case of free vibrations we have $Q = 0$ in eqns (17). The results obtained for different numbers of circumferential waves $\alpha$ are presented in Table 1. Here the natural frequencies calculated by proposed reduced method and finite element method (FEM) are demonstrated. Our results showed good agreement with ones obtained by usage the finite element complex.

Table 1: Natural frequencies of cylindrical tank, $H = 0.8$ m.

| α | $m$ | Natural frequencies, Hz | | |
|---|---|---|---|---|
| | | Empty tank | Fluid-filled tank | |
| | | | Presented method | FEM |
| 0 | 1 | 23.23 | 8.15 | 8.06 |
| | 2 | 91.10 | 45.04 | 44.71 |
| 1 | 1 | 48.52 | 20.98 | 20.86 |
| | 2 | 145.30 | 77.28 | 77.04 |
| 2 | 1 | 79.77 | 40.04 | 39.90 |
| | 2 | 117.07 | 109.51 | 108.89 |

The next example is connected with forced vibrations coupled problem for cylindrical tank practically filled with the fluid. The radial load is suddenly applied to cylindrical surface of the tank $Q(t) = Q_0 T(t)$, where $Q_0 = 10$ MPa – distributed pressure, $T(t) = \exp(-t/\tau)$, $\tau = 14.2 \cdot 10^{-6}$ s.

Figures 2 and 3 are representative of the forced motion response as calculated by proposed method – solid lines and by the finite element complex – dash lines. Figure 2 shows the radial displacement response calculated in point 1 and Figure 3 shows the same one calculated in point 2. The point 1 is situated in the wetted part of the wall whereas point 2 belongs to boundary of the liquid free surface.

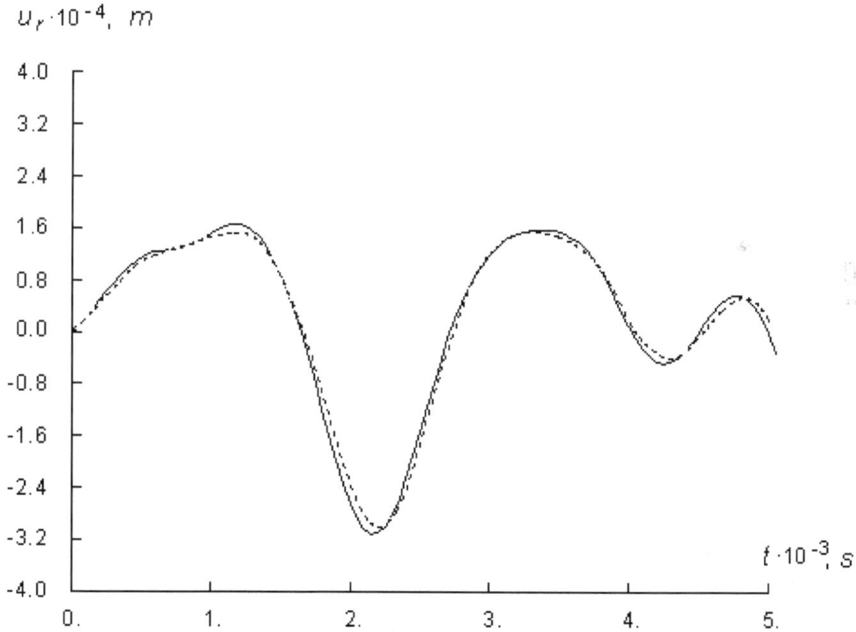


Figure 2: Time history of the radial displacement at point 1.

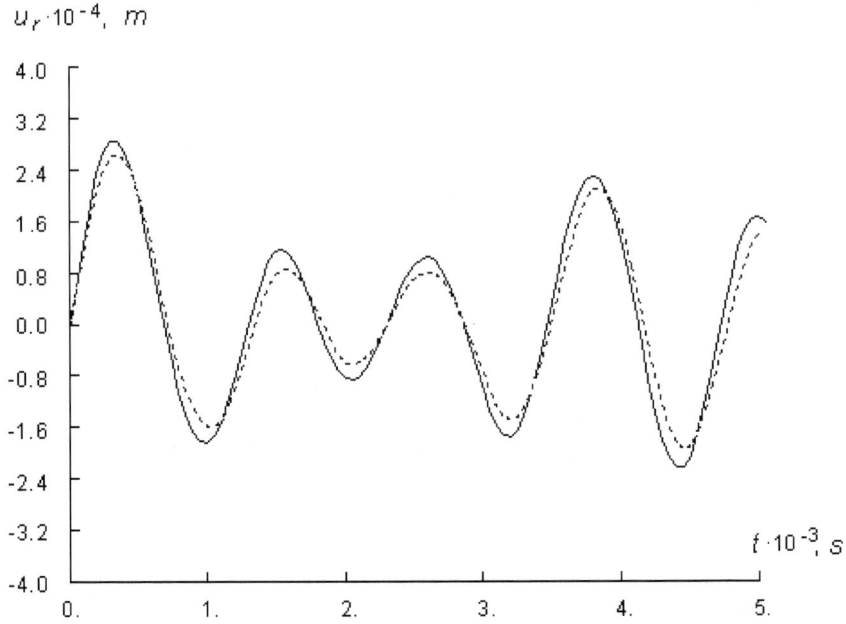


Figure 3: Time history of the radial displacement at point 2.

These figures illustrate the frequency slowness of those shell parts that are interacting with the fluid. Figures also demonstrate good agreement of results obtained by different methods. That testifies the reliability of the method and algorithm proposed. It would be noted that FEM analysis requires 3D modeling to solve this coupled problem.

# 6 Conclusions

The numerical procedure based on a coupling the finite element formulation and the boundary element method is developed for the analysis of free and forced vibrations of shells of revolution with an arbitrary meridian partially filled with the fluid. Integration by the fluid volume is reduced to integrals along the shell meridian and along the radius of the liquid free surface. It is the basic advantage of our method based on a combination of the boundary integral equations method and expansion into Fourier series. The governing integral equations for each harmonic have been obtained. According to the approach the free vibrations problem was reduced to consecutive definition of frequencies and modes of the shell vibrations in vacuum (the first stage), and to the solution of the mixed problems for Laplace equation (the second stage). The forced vibration problem includes the same steps involving the liquid added masses into equations of motion. Numerical investigations of natural frequencies and forced vibrations of the cylindrical tank filled with the incompressible fluid have been carried out.

## Acknowledgements

The authors gratefully acknowledge the sponsorship of this research by the Science and Technology Center in Ukraine in framework of the Project #4624. Authors also would like to acknowledge our STCU Project collaborators Professors Carlos Brebbia and Eduard Ventsel for their constant support and interest to our research.

## References

[1] Kubenko, V.D. & Koval'chuk, P.S., Nonlinear problems of the dynamics of elastic shells partially filled with a liquid. *International Journal of Applied Mechanics*, **36(4)**, pp. 421–448, 2000.

[2] Amabili, M. & Païdoussis, M. P., Review of studies on geometrically nonlinear vibrations and dynamics of circular cylindrical shells and panels, with and without fluid-structure interaction. *Applied Mechanics Review*, **56(4)**, pp. 349–381, 2003.

[3] Kumar, V. & Ganesan, N., Dynamic analysis of conical shells conveying fluid. *Journal of Sound and Vibration*, **310(1-2)**, pp. 38–57, 2008.

[4] Malhotra, P. K., New method for seismic isolation of liquid-storage tanks. *Journal of Earthquake Engineering and Structural Dynamics*, **26(8)**, pp. 839–847, 1997.

[5] Zhang, Y.L., Gorman, D.G. & Reese, J.M., Vibration of prestressed thin cylindrical shells conveying fluid. *Thin-Walled Structures*, **41**, pp. 1103–1127, 2003.

[6] Chen, Y.H., Hwang, W.S. & Ko, C.H., Numerical simulation of the three-dimensional sloshing problem by boundary element method. *Journal of the Chinese Institute of Engineers*, **23(3)**, pp. 321-330, 2000.

[7] Brebbia, C.A., Telles, J.C.F. & Wrobel, L.C. *Boundary Element Techniques*, Springer-Verlag: Berlin and New York, 1984.

# Section 6
# Advanced formulations

# 3-D acoustic shape sensitivity analysis using the fast multipole boundary element method

C. J. Zheng[1,2], T. Matsumoto[1], T. Takahashi[1] & H. B. Chen[2]
[1]*Nagoya University, Japan*
[2]*University of Science and Technology of China, China*

## Abstract

The fast multipole boundary element method (FMBEM), based on the Burton-Miller formulation for 3-D acoustic sensitivity analysis, is presented in this paper in order to overcome the difficulties in the shape sensitivity analyses using the boundary element method based on the direct differentiation method.

The Burton-Miller formulation, which is a linear combination of the conventional boundary integral equation (CBIE) and its normal derivative (NDBIE), is applied to circumvent the difficulty caused by the so-called fictitious eigenfrequencies. The fast multipole method (FMM) is also employed to improve the overall computational efficiency. The sensitivity boundary integral equations of hypersingular type are obtained by the direct differentiation method. The correctness and validity of the method are demonstrated through some numerical results, from which the effectiveness of the present method is shown for 3-D acoustic shape sensitivity analyses.
*Keywords: acoustics, shape sensitivity, direct differentiation method, fictitious eigenfrequency, Burton-Miller's method, fast multipole boundary element method.*

## 1 Introduction

The process of computing gradients of the defined acoustic performance function associated with a structure with respect to design variables has been termed as the acoustic sensitivity analysis. Although the BEM has the incomparable superiority, which makes it widely used in the acoustic sensitivity analysis, it also has some difficulties in solving exterior acoustic problems. The first one is that the BEM with the CBIE fails to yield unique solutions for exterior problems at the eigenfrequencies of the associated interior problems. These eigenfrequencies

WIT Transactions on Modelling and Simulation, Vol 50, © 2010 WIT Press
www.witpress.com, ISSN 1743-355X (on-line)
doi:10.2495/BE100191

are called fictitious eigenfrequencies because they do not have any physical significance. In order to tackle this difficulty, the Burton-Miller method is a more sound and effective method than any other existing ones [1].

The second difficulty, which also restricts the usage of the BEM in large-scale engineering problems, is the high computational complexity. For a problem with $N$ unknowns, a direct solver, such as the Gauss elimination method, requires $O(N^3)$ solution cost and storage of $O(N^2)$. Even worse, in the sensitivity analysis by the direct differentiation method, more integral evaluations for each pair of boundary elements are needed. However, the computation can be performed in $O(N)$ operations and $O(N)$ memory requirement by using the FMBEM, for example, in potential problems. The FMBEM was first proposed by Rokhlin [2], and a comprehensive review can be found in [3].

In this paper, the FMBEM based on the Burton-Miller formulation for 3-D acoustic problems is adopted to make the acoustic shape sensitivity analysis more efficient. The well-known hypersingular nature of the NDBIE can be evaluated without any difficulty by using the constant triangular element discretization, which also makes the FMBEM more efficient than the regularization technique because multipole expansion formulas and other translation formulas only have to be implemented for the fundamental solution and its derivatives of the Helmholtz equation.

## 2 Formulations

### 2.1 Conventional boundary element method (CBEM) formulations

The propagation of time-harmonic acoustic waves in a homogeneous and isotropic acoustic medium is described by the following Helmholtz equation

$$\nabla^2 p(x) + k^2 p(x) = 0, \tag{1}$$

where $p(x)$ is the sound pressure at point $x$, and $k = \omega/c$ the wave number, $\omega$ the circular frequency, $c$ the acoustic velocity.

The boundary conditions for Helmholtz equation are written as

$$p(x) = \bar{p}(x), \qquad \text{on } \Gamma_p, \tag{2}$$

$$q(x) = \frac{\partial p}{\partial n}(x) = i\rho\omega\bar{v}(x), \quad \text{on } \Gamma_q, \tag{3}$$

$$p(x) = zv(x), \quad \text{on } \Gamma_z, \tag{4}$$

where $n(x)$ denotes the unite outward normal vector at point $x$, $i$ the imaginary unit, $\rho$ the medium density, $v(x)$ the particle velocity and $z$ the acoustic impedance. The quantities denoted with bars are assumed to be given values on the boundary. In the case of acoustic radiation, $p(x)$ must also satisfy the Sommerfeld's radiation condition.

### 2.1.1 CBEM formulations for acoustic state analysis

The integral representation of the solution to the Helmholtz equation is

$$p(x) + \int_\Gamma q^*(x,y)p(y)\,\mathrm{d}\Gamma(y) = \int_\Gamma p^*(x,y)q(y)\,\mathrm{d}\Gamma(y). \tag{5}$$

The CBIE and NDBIE of the Helmholtz equation are written, as follows:

$$C(x)p(x) + ⨍_\Gamma q^*(x,y)p(y)\,\mathrm{d}\Gamma(y) = \int_\Gamma p^*(x,y)q(y)\,\mathrm{d}\Gamma(y), \tag{6}$$

$$C(x)q(x) + ⨎_\Gamma \tilde{q}^*(x,y)p(y)\,\mathrm{d}\Gamma(y) = ⨍_\Gamma \tilde{p}^*(x,y)q(y)\,\mathrm{d}\Gamma(y), \tag{7}$$

where $\tilde{( )} = \partial( )/\partial n(x)$, the constant $C(x)$ is $1/2$ if $\Gamma$ is smooth around $x$, and $⨍$ and $⨎$ denotes the integral is evaluated in the sense of Cauchy's principal value and finite part of divergent integral, respectively. $p^*(x,y)$ is the fundamental solution, for 3-D problems it is given as

$$p^*(x,y) = \frac{e^{ikr}}{4\pi r}, \tag{8}$$

where $r = |x-y|$, and $q^*(x,y)$ is the normal derivative of $p^*(x,y)$.

Both eqns (6) and (7) can be used to calculate the unknown boundary state values. For exterior acoustic problems, by using either of them solely, unique solutions cannot be obtained at the fictitious eigenfrequencies. However, Burton and Miller [1] have shown that we can obtain unique solutions for all such frequencies by using a linear combination of eqns (6) and (7) in the form:

$$\begin{aligned} C(x)p(x) &+ ⨍_\Gamma q^*(x,y)p(y)\,\mathrm{d}\Gamma(y) + \alpha ⨎_\Gamma \tilde{q}^*(x,y)p(y)\,\mathrm{d}\Gamma(y) \\ &= -\alpha C(x)q(x) + \int_\Gamma p^*(x,y)q(y)\,\mathrm{d}\Gamma(y) + \alpha ⨍_\Gamma \tilde{p}^*(x,y)q(y)\,\mathrm{d}\Gamma(y), \end{aligned} \tag{9}$$

where $\alpha$ is a coupling constant that can be chosen as $i/k$ [4].

Discretizing eqn (9), collecting the equations for all the collocation points and expressing them in matrix forms result in the following form of linear algebraic equations:

$$[H]\{p\} = [G]\{q\}. \tag{10}$$

Rearranging eqn (10) gives the following system of linear equations

$$[A]\{x\} = [B]\{y\}, \tag{11}$$

where $[A]$ is the system matrix, $\{x\}$ and $\{y\}$ are the unknown and known vectors.

Eqn (11) can now be solved and all the boundary state values are then known. After that, it is possible to calculate $p(x)$ at any point $x$ within the problem domain by using eqn (5).

### 2.1.2 CBEM formulations for acoustic sensitivity analysis

In order to obtain the formulations for sensitivity analysis, eqns (5), (6) and (7) are differentiated with respect to an arbitrary shape design variable to yield

$$\begin{aligned}&\dot{p}(x) + \int_\Gamma q^*(x,y)\dot{p}(y)\,\mathrm{d}\Gamma(y) + \int_\Gamma \dot{q}^*(x,y)p(y)\,\mathrm{d}\Gamma(y)\\ &+ \int_\Gamma q^*(x,y)p(y)\,\mathrm{d}\dot{\Gamma}(y) = \int_\Gamma p^*(x,y)\dot{q}(y)\,\mathrm{d}\Gamma(y)\\ &+ \int_\Gamma \dot{p}^*(x,y)q(y)\,\mathrm{d}\Gamma(y) + \int_\Gamma p^*(x,y)q(y)\,\mathrm{d}\dot{\Gamma}(y),\end{aligned} \tag{12}$$

$$\begin{aligned}&C(x)\dot{p}(x) + \fint_\Gamma q^*(x,y)\dot{p}(y)\,\mathrm{d}\Gamma(y) + \fint_\Gamma \dot{q}^*(x,y)p(y)\,\mathrm{d}\Gamma(y)\\ &+ \fint_\Gamma q^*(x,y)p(y)\,\mathrm{d}\dot{\Gamma}(y) = \int_\Gamma p^*(x,y)\dot{q}(y)\,\mathrm{d}\Gamma(y)\\ &+ \int_\Gamma \dot{p}^*(x,y)q(y)\,\mathrm{d}\Gamma(y) + \int_\Gamma p^*(x,y)q(y)\,\mathrm{d}\dot{\Gamma}(y),\end{aligned} \tag{13}$$

$$\begin{aligned}&C(x)\dot{q}(x) + \int\!\!\!\!\!=_\Gamma \tilde{q}^*(x,y)\dot{p}(y)\,\mathrm{d}\Gamma(y) + \int\!\!\!\!\!=_\Gamma \dot{\tilde{q}}^*(x,y)p(y)\,\mathrm{d}\Gamma(y)\\ &+ \int\!\!\!\!\!=_\Gamma \tilde{q}^*(x,y)p(y)\,\mathrm{d}\dot{\Gamma}(y) = \fint_\Gamma \tilde{p}^*(x,y)\dot{q}(y)\,\mathrm{d}\Gamma(y)\\ &+ \fint_\Gamma \dot{\tilde{p}}^*(x,y)q(y)\,\mathrm{d}\Gamma(y) + \fint_\Gamma \tilde{p}^*(x,y)q(y)\,\mathrm{d}\dot{\Gamma}(y),\end{aligned} \tag{14}$$

where the upper dot $\dot{(\;)}$ denotes differentiation with respect to the design variable, and $\dot{p}^*(x,y)$ can be related to the sensitivities of the coordinate as

$$\dot{p}^*(x,y) = -\frac{e^{ikr}}{4\pi r^2}(1-ikr)\frac{\partial r}{\partial y_i}(\dot{y}_i - \dot{x}_i), \tag{15}$$

and $\dot{q}^*(x,y)$, $\dot{\tilde{p}}^*(x,y)$ and $\dot{\tilde{p}}^*(x,y)$ can also be expressed in the form as $\dot{p}^*(x,y)$.

Both eqns (13) and (14) can be used for acoustic boundary sensitivity analysis. But for an exterior acoustic problem, similar to eqns (6) and (7), they also have a different set of fictitious eigenfrequencies, and the linear combination of them can also provide unique solutions for all frequencies.

Discretizing the linearly composite formulation of eqns (13) and (14), and collecting the equations for all the collocation points and writing them in matrix forms result in the following form of linear algebraic equations

$$[H]\{\dot{p}\} = [G]\{\dot{q}\} - [h]\{p\} + [g]\{q\}, \tag{16}$$

where $[H]$ and $[G]$ are the same matrices as those in eqn (10), but $[h]$ and $[g]$ are the newly derived matrices that must be computed here. Also, the unknown terms of $\{p\}$ and $\{q\}$ in eqn (16) can be obtained by solving eqn (11).

$\dot{p}$ (or $\dot{q}$) is also known on the boundary where $p$ (or $q$) is prescribed. Therefore, eqn (16) can be rearranged by using the boundary conditions for $\dot{p}$ and $\dot{q}$ as follows

$$[A]\{\dot{x}\} = [B]\{\dot{y}\} - [h]\{p\} + [g]\{q\}, \tag{17}$$

where $[A]$ and $[B]$ are the same matrices as those in eqn (11), $\{\dot{x}\}$ and $\{\dot{y}\}$ are the unknown and known vectors, respectively. Equation (17) can now be solved and all the boundary sensitivity values are then known. Once this has been done, it is possible to calculate $\dot{p}(x)$ at any point within the problem domain by using eqn (12).

The strongly-singular or hypersingular boundary integrals can be found in eqns (6), (7), (13) and (14). In order to calculate them accurately, eqns (6), (7), (13) and (14) are often regularized by using the fundamental solution of the Laplace's equation. But the regularized equations are not efficient to use in the FMBEM, because multipole expansion formulas and other translation formulas have to be implemented not only for the fundamental solution and its derivatives of the Helmholtz's equation but also for those of the Laplace's equation. However, when the constant triangular element is used to discretize the boundary, these strongly singular and hypersingular integrals can be evaluated in the sense of finite parts without any difficulty, hence FMBEM can be applied efficiently. The details of evaluating such integrals for constant triangular elements can be found in [5, 6].

## 2.2 FMBEM formulations

In this section, the FMBEM is employed to solve the problems, and an iterative solver, the generalized minimal residual method (GMRES), is used in the solution process because the system of linear equations (11) and (17) is not formed explicitly. Several expansion and translation formulas needed in the procedure of the FMBEM have been well presented in [7].

The fundamental solution (8) can be expanded into the following series around an expansion point $O$ near $y$

$$p^*(x,y) = \frac{ik}{4\pi}\sum_{n=0}^{\infty}\sum_{m=-n}^{n}(2n+1)\bar{I}_n^m(k,\overrightarrow{Oy})O_n^m(k,\overrightarrow{Ox}), \quad |\overrightarrow{Ox}| > |\overrightarrow{Oy}|, \tag{18}$$

where

$$I_n^m(k,\overrightarrow{Oy}) = j_n(k|\overrightarrow{Oy}|)Y_n^m(\widehat{Oy}), \tag{19}$$

$$O_n^m(k,\overrightarrow{Ox}) = h_n^{(1)}(k|\overrightarrow{Ox}|)Y_n^m(\widehat{Ox}), \tag{20}$$

and $\bar{I}_n^m$ is the complex conjugate of $I_n^m$, $j_n$ and $h_n^{(1)}$ are the $n$-th order spherical Bessel function of the first and third kind, and $Y_n^m$ is the spherical harmonics.

### 2.2.1 FMBEM formulations for acoustic state analysis

In the FMBEM procedure, we calculate first the following multipole moments for the boundary element $j$ far away from the collocation point $x$ ($|\overrightarrow{Ox}| > |\overrightarrow{Oy_j}|$).

$$M_n^m(k, \overrightarrow{Oy_j}) = \begin{cases} \displaystyle\int_{\Gamma_j} \frac{\partial \bar{I}_n^m(k, \overrightarrow{Oy})}{\partial n(y)} p(y)\, \mathrm{d}\Gamma(y) & \text{for } H^{ij} p^j, \\ \displaystyle\int_{\Gamma_j} \bar{I}_n^m(k, \overrightarrow{Oy}) q(y)\, \mathrm{d}\Gamma(y) & \text{for } G^{ij} q^j. \end{cases} \tag{21}$$

Multipole moments of a group of $\ell$ elements that are close to $O$ (more precisely, all elements in the same LEAF-cell whose center is point $O$) can be added up in order to form $M_n^m(k, O)$ that is the multipole moments centered at $O$.

$$M_n^m(k, O) = \sum_{j=1}^{\ell} M_n^m(k, \overrightarrow{Oy_j}). \tag{22}$$

The center of the multipole moment can be shifted from $O$ to $O'$ according to the following Moments to Moments (M2M) translation formulation, if $|\overrightarrow{O'x}| > |\overrightarrow{O'y}|$.

$$M_n^m(k, O') = \sum_{n'=0}^{\infty} \sum_{m'=-n'}^{n'} \sum_{\substack{l=|n-n'| \\ n'+n-l:even}}^{n+n'} (2n'+1)(-1)^{m'} W_{n,n',m,m',l} \times I_l^{-m-m'}(k, \overrightarrow{O'O}) M_{n'}^{-m'}(k, O), \tag{23}$$

where $W_{n,n',m,m',l}$ is given by

$$W_{n,n',m,m',l} = (2l+1) i^{n'-n+l} \begin{pmatrix} n & n' & l \\ 0 & 0 & 0 \end{pmatrix} \begin{pmatrix} n & n' & l \\ m & m' & -m-m' \end{pmatrix}, \tag{24}$$

and $\begin{pmatrix} \cdot & \cdot & \cdot \\ \cdot & \cdot & \cdot \end{pmatrix}$ stands for the Wigner 3j symbol [8].

The Moments to Local expansion (M2L) translation can be expressed as

$$L_n^m(k, x_0) = \sum_{n'=0}^{\infty} \sum_{m'=-n'}^{n'} \sum_{\substack{l=|n-n'| \\ n'+n-l:even}}^{n+n'} (2n'+1)(-1)^{m+m'} W_{n',n,m',m,l} \times O_l^{m+m'}(k, \overrightarrow{O'x_0}) M_{n'}^{m'}(k, O'). \tag{25}$$

The local expansion center can be shifted from $x_0$ to $x_1$ according to the following Local expansion to Local expansion (L2L) translation formulation,

$$L_n^m(k,x_1) = \sum_{n'=0}^{\infty} \sum_{m'=-n'}^{n'} \sum_{\substack{l=|n-n'| \\ n'+n-l:even}}^{n+n'} (2n'+1)(-1)^m W_{n',n,m',-m,l} \times I_l^{m-m'}(k,\overrightarrow{x_0x_1}) L_{n'}^{m'}(k,x_0). \quad (26)$$

Finally, $\sum_{j=1}^{\ell} H^{ij}p^j$ or $\sum_{j=1}^{\ell} G^{ij}q^j$ can be expressed in terms of the local expansions using local expansion coefficients obtained from eqns (25) or (26) as

$$\sum_{j=1}^{\ell} H^{ij}p^j \text{ or } \sum_{j=1}^{\ell} G^{ij}q^j = \frac{ik}{4\pi} \sum_{n=0}^{\infty} \sum_{m=-n}^{n} (2n+1) L_n^m(k,x_1) \times \left( \bar{I}_n^m(k,\overrightarrow{x_1x}) + \alpha \frac{\partial \bar{I}_n^m(k,\overrightarrow{x_1x})}{\partial n(x)} \right). \quad (27)$$

For elements that are close to the collocation point $x$, the conventional method in Section 2.1.1 should be used.

### 2.2.2 FMBEM formulations for acoustic sensitivity analysis

The computation of $\sum_{j=1}^{\ell} H^{ij}\dot{p}^j$ or $\sum_{j=1}^{\ell} G^{ij}\dot{q}^j$ by using the FMBEM is similar to that of $\sum_{j=1}^{\ell} H^{ij}p^j$ or $\sum_{j=1}^{\ell} G^{ij}q^j$ in the above section, so only the computation of $\sum_{j=1}^{\ell}(g^{ij}q^j - h^{ij}p^j)$ is discussed next.

It is easy to find that the computation of $\sum_{j=1}^{\ell}(g^{ij}q^j - h^{ij}p^j)$ should be done in two steps, the multipole moments for these two steps are defined, as follows:

$${}^1M_n^m(k,\overrightarrow{Oy_j}) = \int_{\Gamma_j} \bar{I}_n^m(k,\overrightarrow{Oy}) q(y)\, \mathrm{d}\Gamma(y) - \int_{\Gamma_j} \frac{\partial \bar{I}_n^m(k,\overrightarrow{Oy})}{\partial n(y)} p(y)\, \mathrm{d}\Gamma(y), \quad (28)$$

and

$${}^2M_n^m(k,\overrightarrow{Oy_j}) = \int_{\Gamma_j} \left( \frac{\partial \bar{I}_n^m(k,\overrightarrow{Oy})}{\partial y_i} \dot{y}_i q^j \right) \mathrm{d}\Gamma(y) + \int_{\Gamma_j} \bar{I}_n^m(k,\overrightarrow{Oy}) q^j\, \dot{\mathrm{d}\Gamma}(y)$$
$$- \int_{\Gamma_j} \left( \frac{\partial^2 \bar{I}_n^m(k,\overrightarrow{Oy})}{\partial y_i \partial y_j} n_i(y) \dot{y}_j p^j \right) \mathrm{d}\Gamma(y) - \int_{\Gamma_j} \left( \frac{\partial \bar{I}_n^m(k,\overrightarrow{Oy})}{\partial y_i} \dot{n}_i(y) p^j \right) \mathrm{d}\Gamma(y)$$
$$- \int_{\Gamma_j} \left( \frac{\partial \bar{I}_n^m(k,\overrightarrow{Oy})}{\partial y_i} n_i(y) p^j \right) \dot{\mathrm{d}\Gamma}(y). \quad (29)$$

The M2M, M2L, L2L translation formulations for the two steps are the same as eqns (23), (25) and (26). Finally, $\sum_{j=1}^{\ell}(g^{ij}q^j - h^{ij}p^j)$ can be expressed in terms of the local expansions as

$$\sum_{j=1}^{\ell}(g^{ij}q^j - h^{ij}p^j) = \frac{ik}{4\pi}\sum_{n=0}^{\infty}\sum_{m=-n}^{n}(2n+1)\,{}^1L_n^m(k,x_1)\left(\frac{\partial \bar{I}_n^m(k,\overrightarrow{x_1x})}{\partial x_i}\dot{x}_i\right.$$
$$\left. + \alpha\frac{\partial^2 \bar{I}_n^m(k,\overrightarrow{x_1x})}{\partial x_i \partial x_j}n_i(x)\dot{x}_j + \alpha\frac{\partial \bar{I}_n^m(k,\overrightarrow{x_1x})}{\partial x_i}\dot{n}_i(x)\right)$$
$$+ \frac{ik}{4\pi}\sum_{n=0}^{\infty}\sum_{m=-n}^{n}(2n+1)\,{}^2L_n^m(k,x_1)\left(\bar{I}_n^m(k,\overrightarrow{x_1x}) + \alpha\frac{\partial \bar{I}_n^m(k,\overrightarrow{x_1x})}{\partial n(x)}\right). \quad (30)$$

For elements that are close to the collocation point $x$, the integrals are evaluated with the conventional method stated in Section 2.1.2.

## 3 Numerical examples

A pulsating sphere example that has the analytical solution is chosen to verify the validity of the method. The radius $a$ of the sphere is designated as a shape design variable. The partial derivative of the sound pressure at a field point with respect to this radius is calculated and compared with the analytical solution.

For a sphere vibrating with the velocity $v_n$ in the radial direction, the sensitivity of the pressure with respect to a change of radius becomes

$$\frac{\partial p(r)}{\partial a} = \frac{\rho cka}{r}\left\{\frac{\partial v_n}{\partial a}\frac{ia}{1+ika} + v_n\left[\frac{2i-ka}{1+ika} + \frac{ka}{(1+ika)^2}\right]\right\}e^{-ik(r-a)}, \quad (31)$$

where $r$ is a distance from the center of the sphere to the field point of interest.

In the numerical analysis, we assumed as $a = 1.0$m, $v_n = 1.0$m/s, and sensitivities are calculated for a field point with $r = 6.0$m. The acoustic medium is assumed to be air with a density of $\rho = 1.2\,\mathrm{kg/m^3}$ and a sound velocity of $c = 340\,\mathrm{m/s}$. The values of $\dot{x}_i$ used for eqns (29) and (30) were given, for simplicity, as the finite difference approximations calculated from the radius augmented by 1%.

Figures 1 and 2 show the result for a 1078 constant triangular element model with the wave numbers varying from 0.5 to 5.0. The solutions of real and imaginary part of the sound pressure sensitivities are shown. It can be seen that the CBIE or NDBIE solution follows the analytical solution closely except in the vicinity of the fictitious eigenfrequencies. But the solution based on Burton-Miller's Method is very accurate over the entire range of the wave number.

Figures 3 and 4 show comparisons of relative errors and CPU times between FMBEM, CBEM-LU and CBEM-GMRES. In all the computations, the maximum number of elements in a leaf is set to 100, and the truncation number of multipole and local expansion terms is set to10. The GMRES solver is to stop iterations when the residual becomes smaller than the tolerance $10^{-3}$, and $k$ is set to 0.5. All computations were done on a PC with an Intel 3.0GHz processor and 2GB memory. Figure 3 shows that the percentage relative errors of the FMBEM,

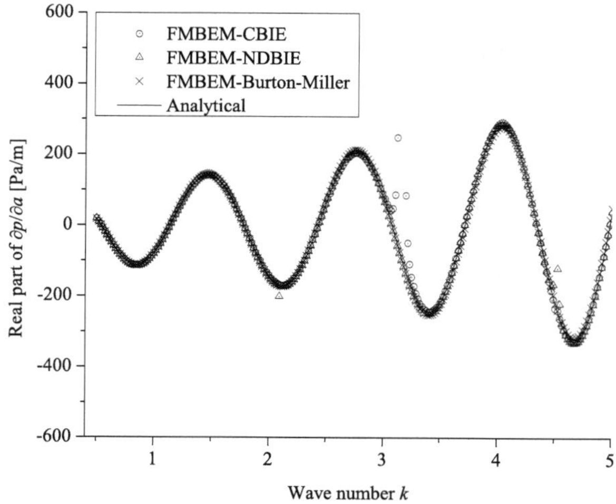


Figure 1: Results for real part of $\partial p(r)/\partial a$.

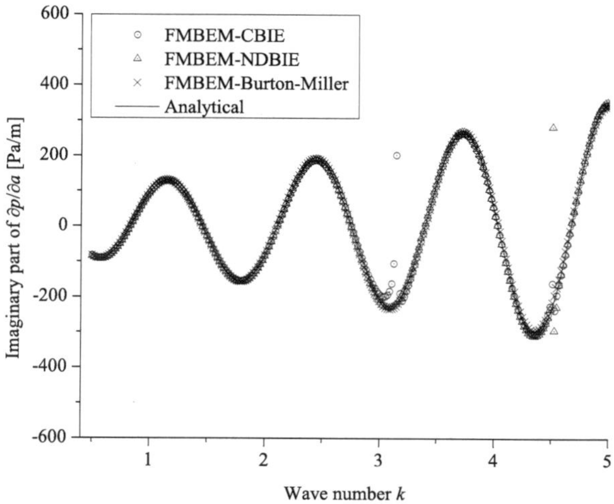


Figure 2: Results for imaginary part of $\partial p(r)/\partial a$.

CBEM-LU and CBEM-GMRES are very close to each other, and all errors decrease very fast as the number of elements increases. It is found from Figure 4 that the FMBEM is faster than the CBEM-LU for models with more than 4000 elements, and the FMBEM is also faster than the CBEM-GMRES for models with more than 7000 elements.

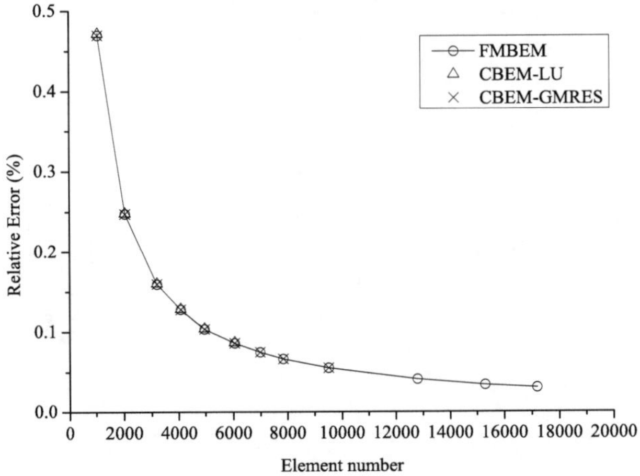


Figure 3: Comparison of relative error.

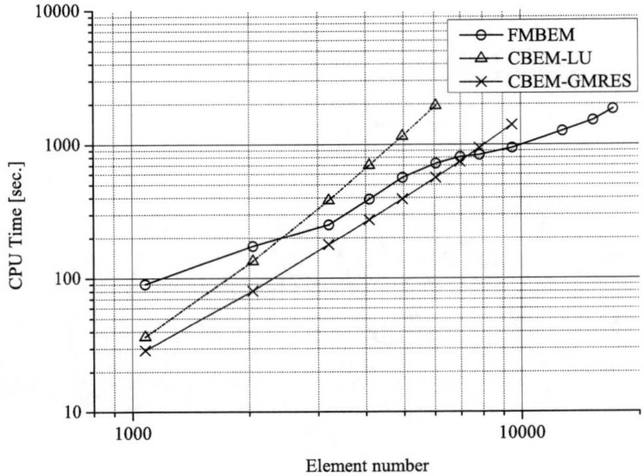


Figure 4: Comparison of CPU time.

## 4 Conclusions

A FMBEM approach has been presented in this paper for 3-D acoustic design sensitivity analysis using the boundary integral equation of shape design sensitivity coefficients of the sound pressure and particle velocity. Since the so-called fictitious eigenfrequencies are also observed for the sensitivity analysis for exterior acoustic fields, the Burton-Miller approach, which uses a linear combination of CBIE and NDBIE for the sensitivity coefficients, is employed to overcome the problem.

The multipole moments and M2M, M2L, L2L translation formulas have been presented for the derived sensitivity boundary integral equations. Some numerical results for a pulsating sphere has been shown to demonstrate the validity and computation efficiency of the present approach.

## References


[1] Burton, A.J. & Miller, G.F., The application of integral equation methods to the numerical solution of some exterior boundary-value problems. *Proc. Roy. Soc. Lond.*, A. **323**, pp. 201-210, 1971.

[2] Rokhlin, V., Rapid solution of integral equations of classical potential theory. *Journal of Computational Physics*, **60(2)**, pp. 187-207, 1985.

[3] Nishimura, N., Fast multipole accelerated boundary integral equation methods. *ASME*, **55(4)**, pp. 299-324, 2002.

[4] Kress, R., Minimizing the condition number of bounday integral operators in acoustic and electromagnetic scattering. *Q. J. Mech Appl Math*, **38**, pp. 323-243, 1985.

[5] Matsumoto, T., Zheng, C.J., Harada, S. & Takahashi, T., Explicit evaluation of hypersingular boundary integral equation of 3-D Helmholtz equation for constant triangular element discretization. Submitted to JSME Technical Journal.

[6] Zheng, C.J., Matsumoto, T., Takahashi, T. & Chen, H.B., Boundary element shape design sensitivity formulation of 3D acoustic problems based on direct differentiation of strongly-singular and hypersingular boundary integral equations. Submitted to JSME.

[7] Yoshida, K., PhD Thesis, Kyoto University, Japan, 2001.

[8] Abramowitz, M. & Stegun, I.A., *Handbook of Mathematical Functions with Formulas*, Dover: New York, pp. 1006-1010, 1972.

[9] Matsumoto, T., Tanaka, M. & Yamada, Y., Design Sensitivity Analysis of Steady-State Acoustic Problems using Boundary Integral Equation Formulation. *JSME International Journal, Series C*, **38(1)**, pp. 9-16, 1995.

[10] Arai, Y., Tanaka, M. & Matsumoto, T., A New Design Sensitivity Analysis of Acoustic Problems Based on Bem Avoiding Fictitious Eigenfrequency Issure. (in Japanese) *JSME, Series C*, **73(729)**, pp. 120-127, 2007.

[11] Shen, L. & Liu, Y.J., An adaptive fast multipole boundary element method for three-dimensional acoustic wave problems based on the Burton-Miller formulation, *Comput Mech*, **40**, pp. 461-472, 2007.

# The boundary element method revisited

N. A. Dumont
*Civil Engineering Department,*
*Pontifical Catholic University of Rio de Janeiro, Brazil*

## Abstract

The collocation boundary element method is derived on the basis of the weighted-residuals statement. Only the static case is addressed, as it already involves all relevant conceptual issues. The present outline brings to discussion some relevant aspects and implementation issues of the method that should belong in any text book. It is shown that, if the boundary element method is consistently formulated, an inherent error term – related to arbitrary rigid-body displacements – is naturally taken into account and has no influence on the resultant matrix equation, with traction force parameters that are always in balance independently of mesh discretization. The constitutive matrices of the method – the single-layer and double-layer potential matrices **G** and **H** – present some spectral properties that are per se interesting but that also have applicability consequences. The matrix **G** is rectangular, if consistently obtained. For and adequately formulated problem, the solution of the resultant matrix equation is always possible (and unique) whether directly or approximated in terms of equivalent nodal forces. The effects of body forces, whenever transformable to boundary actions, may be expressed in terms of the boundary interpolation functions, which renders the final matrix equation more elegant and speeds up calculations in no detriment to accuracy. There is a novel proposition for the interpolation of traction forces along curved boundaries, with results that may be only slightly improved, as compared to the classical procedure, but that simplifies numerical computation and adds to the consistency of the method in terms of patch test assessments. The conceptual and numerical developments are illustrated by means of a few examples.

*Keywords: boundary elements, weighted-residual methods, generalized inverses.*

WIT Transactions on Modelling and Simulation, Vol 50, © 2010 WIT Press
www.witpress.com, ISSN 1743-355X (on-line)
doi:10.2495/BE100201

# 1 Introduction

The conventional, collocation boundary element method (CBEM), whenever applicable, is a simple, powerful numerical analysis tool [1]. There is an uncountable number of alternatives of formulations and implementations, some of them based on sound principles [7, 8], many of them contributing to the richness of the method [1], but unfortunately some of them just misleading. The present contribution is not a review paper and, on the contrary, makes just a few references to the technical literature. In this outline, which starts with a consistent derivation of the CBEM, one pays tribute to the method by addressing some theoretical issues that are in general overlooked. One expects to add to the conceptual consolidation of the method by analyzing some relevant spectral properties, and also daring to suggest a few improvements.

# 2 A simple, consistent derivation of the BEM

## 2.1 Problem formulation

An elastic body is submitted to body forces $b_i$ in the domain $\Omega$ and traction forces $\bar{t}_i$ on part $\Gamma_\sigma$ of the boundary. Displacements $\bar{u}_i$ are known on the complementary part $\Gamma_u$ of $\Gamma$. One is looking for an adequate approximation of the stress field that satisfies equilibrium in the domain,

$$\sigma_{ji},_j + b_i = 0 \quad \text{in } \Omega, \tag{1}$$

also satisfying the boundary equilibrium and compatibility equations,

$$\sigma_{ji}\eta_j = \bar{t}_i \quad \text{along } \Gamma_\sigma, \quad u_i = \bar{u}_i \quad \text{on } \Gamma_u, \tag{2}$$

where $\eta_j$ is the outward unit normal to $\Gamma$. Indices *i, j,* (also *k, l*) may assume values 1, 2 or 3, as they refer to the coordinate directions *x*, *y* or *z*, respectively, for a general 3D analysis. Sum is indicated by repeated indices. Particularization to 2D analysis as well as to potential problems is straightforward.

## 2.2 From a variational to a consistent weighted-residuals statement

Assuming that $\sigma_{ij}$ is a symmetric tensor that satisfies a priori the constitutive equation $\sigma_{ij} = C_{ijkl} u_{k,l}$, the present problem might be formulated in the frame of the strong form of the principle of stationary (minimum) total potential energy [6], for a variation $\delta u_i$ of $u_i$, already extending the boundary integral from $\Gamma_\sigma$ to $\Gamma$, since, according to eqn (2), $\delta u_i = 0$ on $\Gamma_u$:

$$\delta\Pi = -\int_\Omega \left(\sigma_{ji,j} + b_i\right)\delta u_i \mathrm{d}\Omega + \int_\Gamma \left(\sigma_{ji}\eta_j - \bar{t}_i\right)\delta u_i \mathrm{d}\Gamma = 0 \,. \tag{3}$$

However, one rather formulates the problem in a non-variational, less restrictive framework than eqn (3), in terms of weighted residuals, resorting to a field of *fundamental solutions*, that is, stresses and displacements of the same elasticity problem, $\delta\sigma^*_{ij} = C_{ijkl}\,\delta u^*_{k,l}$, that satisfy the homogeneous part of eqn (1), but not the boundary conditions of eqn (2):

$$-\int_\Omega \left(\sigma_{ji,j} + b_i\right)\delta u_i^* \mathrm{d}\Omega + \int_\Gamma \left(\sigma_{ji}\eta_j - t_i\right)\delta u_i^* \mathrm{d}\Gamma = 0\,. \tag{4}$$

Integrating by parts twice the first term on the left of the above equation, with successive application of the Green's theorem, one obtains

$$\int_\Gamma \delta\sigma^*_{ji}\eta_j u_i \mathrm{d}\Gamma - \int_\Omega \delta\sigma^*_{ji,j} u_i \mathrm{d}\Omega = \int_\Gamma t_i \delta u_i^* \mathrm{d}\Gamma + \int_\Omega b_i \delta u_i^* \mathrm{d}\Omega\,. \tag{5}$$

To arrive at the above expression, one has also resorted to the identity $\sigma_{ji}\delta u^*_{i,j} \equiv u_{k,l}C_{ijkl}\,\delta u^*_{i,j} \equiv u_{k,l}\,\delta\sigma^*_{kl}$, which is not Betti's reciprocity theorem.

The conventional, collocation boundary element method may be derived from eqn (5), for fundamental solutions $\delta\sigma^*_{ij}$ and $\delta u^*_i$ given as

$$\delta\sigma^*_{ij} \equiv \sigma^*_{ijm}\delta p^*_m\,, \tag{6}$$

$$\delta u^*_i = (u^*_{im} + u^r_{is}C_{sm})\delta p^*_m\,, \tag{7}$$

where $u^r_{is}$, for $s = 1\ldots n^r$, are $n^r$ rigid-body displacements that are multiplied by in principle arbitrary constants $C_{sm}$, and $\delta p^*_m$ are arbitrary (virtual) force parameters, with $m$ characterizing both location and direction of application of $\delta p^*_m$. Then, $\delta\sigma^*_{ijm}$ and $\delta u^*_{im}$ are functions – with global support – of the coordinates and directions of $\delta p^*_m$ referred to by $m$ (the source point), as well as of the coordinates and directions referred to by $i$ (the field point), where the effects of $\delta p^*_m$ are measured.

The strength of the boundary element method stems from the fact that $\delta p^*_m$ are point forces applied along $\Gamma$, just outside $\Omega$, however infinitely close. Although $\delta\sigma^*_{ijm}$ and $\delta u^*_{im}$ tend to infinity at the point of application of $\delta p^*_m$, they are analytical in $\Omega$. For convenience, the functions $\delta\sigma^*_{ijm}$ are normalized such that, for a domain $\Omega_0$ that contains $\delta p^*_m$, with enclosing boundary $\Gamma_0$,

$$\int_{\Omega 0} \sigma^*_{jim,j} \mathrm{d}\Omega = \int_{\Gamma 0} \sigma^*_{jim}\eta_j \mathrm{d}\Gamma \equiv -\delta_{im}\,, \tag{8}$$

where $\delta_{im}$ is a generalized Kronecker delta (equal to either 1, when $i$ and $m$ refer to the same degree of freedom, or 0, otherwise). According to the above definition of fundamental solution, the domain integral on the left-hand side of eqn (5) is actually evaluated as $\int_\Omega \delta\sigma^*_{ji,j} u_i \mathrm{d}\Omega = -\delta_{im}u_i\delta p^*_m \equiv -u_m\delta p^*_m$.

Substituting for $\delta\sigma^*_{ijm}$ and $\delta u^*_{im}$ in eqn (5) according to their expressions in eqns (6, 7), one obtains the modified expression of the Somigliana's identity,

$$u_{im} = \int_\Gamma t_i u^*_{im} \mathrm{d}\Gamma - \int_\Gamma \sigma^*_{jim} \eta_j u_i \mathrm{d}\Gamma + \int_\Omega b_i u^*_{im} \mathrm{d}\Omega + C_{sm}\left( \int_\Gamma t_i u^r_{is} \mathrm{d}\Gamma + \int_\Omega b_i u^r_{is} \mathrm{d}\Omega \right), \quad (9)$$

which is used to evaluate displacements $u_{im}$ (and, subsequently, stresses) at a domain point $m$ for prescribed forces $b_i$, $t_i$, and boundary displacements $u_i$. The term in brackets vanishes only if $b_i$ and $t_i$ are in equilibrium, which is not necessarily true when one is dealing with approximations. Then, the results are in principle influenced by some arbitrary constants $C_{sm}$ [1].

### 2.3 Numerical discretization

Equation (9) is also used to evaluate the displacements $u_i$ and the traction forces $t_i$ as the problem's unknowns along $\Gamma_\sigma$ and $\Gamma_u$, respectively. In fact, they may be approximated along $\Gamma$ as

$$u_i = u_{in} d_n, \quad (10)$$

$$t_i = t_{i\ell} t_\ell, \quad (11)$$

where $d_n$, for $n = 1 \ldots n^d$, is a vector of $n^d$ nodal displacements and $u_{in}$ are interpolation functions with local support, usually polynomials chosen in such a way that, at the nodal points, $u_{in} \equiv \delta_{in}$. Since the traction forces $t_i$ are surface attributes, the $n^t$ parameters $t_\ell$ are also surface attributes that depend on the outward normal $\eta_i$ of the boundary point at which $t_\ell$ is physically attached. Generally, $n^t > n^d$, as the boundary $\Gamma$ may not be entirely smooth, with more than one normal at some points. The interpolation functions $t_{i\ell}$ also have local support, but are not necessarily locally expressed as $u_{in}$ (see Section 3).

The boundary geometry is approximated from the nodal attributes using the same interpolation functions $u_{in}$ of eqn (10), which consists in an isoparametric representation of the problem, exactly as in the finite element method.

Replacing $u_i$ and $t_i$ in eqn (9) with their boundary approximations given by eqns (10, 11), and applying $\delta p^*_m$ at successive boundary points in such a way that $\delta p^*_m d_m$ has the meaning of virtual work, one arrives at the basic equation of the conventional, collocation boundary element method, as given in the literature,

$$\left( \int_\Gamma \sigma^*_{jim} \eta_j u_{in} \mathrm{d}\Gamma + \delta_{mn} \right) d_n = \int_\Gamma t_{i\ell} u^*_{im} \mathrm{d}\Gamma t_\ell + \int_\Omega b_i u^*_{im} \mathrm{d}\Omega + C_{sm}\left( \int_\Gamma t_{i\ell} u^r_{is} \mathrm{d}\Gamma t_\ell + \int_\Omega b_i u^r_{is} \mathrm{d}\Omega \right), \quad (12)$$

except for the term related to the constants $C_{sm}$, that is actually an error term. Equation (12) may be written in matrix format as

$$\mathbf{Hd} = \mathbf{Gt} + \mathbf{b} + \boldsymbol{\varepsilon}\,, \tag{13}$$

where $\mathbf{H} = \left[H_{mn}\right] \in R^{n^d \times n^d}$ is a kinematic transformation matrix, $\mathbf{G} = \left[G_{m\ell}\right] \in R^{n^d \times n^t}$ is a flexibility-type matrix and $\mathbf{b} = \left[b_m\right] \in R^{n^d}$ is a vector of nodal displacements equivalent to the applied body forces. The double-layer and single-layer potential matrices $\mathbf{H}$ and $\mathbf{G}$ comprise in their definition singular and improper integrals, respectively, when source (index *m*) and field (index either $n$ or $\ell$) refer to the same nodal points. Then, special care must be taken in the numerical integrations. This has been sufficiently investigated and is not the subject of the present outline. The error term $\boldsymbol{\varepsilon}$ in eqn (13), defined according to eqn (12), corresponds to residuals whose magnitude depends on the amount of rigid-body displacements that are always implicit in the fundamental solution, eqn (7), as well as on how refined is the boundary mesh, that is, how accurately the boundary traction forces, approximated according to eqn (11), are in equilibrium with the applied domain forces $b_i$. This vector of residuals is usually disregarded in the implementations shown in the literature [1, 2], or sometimes used as a measure of convergence of the numerical model. A consistent numerical model must take this term explicitly into account and end up with a formulation that is independent of $C_{sm}$ instead of just disregarding it.

This specific issue has already been the subject of a theoretical investigation [3]. The main results are summarized in the following, also introducing a convenient simplification related to the particular solution term $\mathbf{b}$ of eqn (13).

The vector of residuals $\boldsymbol{\varepsilon}$ in eqn (13) may be written as

$$\boldsymbol{\varepsilon} = \mathbf{C}^{\mathrm{T}}\mathbf{R}^{\mathrm{T}}\left(\mathbf{t} - \mathbf{t}^{p}\right), \tag{14}$$

where $\mathbf{R} = \left[R_{\ell s}\right] \in R^{n^t \times n^r}$ is defined as

$$R_{\ell s} = \int_{\Gamma} t_{i\ell} u^{r}_{is} \mathrm{d}\Gamma\,, \tag{15}$$

and the product $\mathbf{R}^{\mathrm{T}}\mathbf{t}^{p}$ comes from the approximation

$$\int_{\Omega} b_i u^{r}_{is} \mathrm{d}\Omega = -\int_{\Gamma} \sigma^{p}_{ji} \eta_j u^{r}_{is} \mathrm{d}\Gamma \approx -\int_{\Gamma} t_{i\ell} u^{r}_{is} \mathrm{d}\Gamma t^{p}_{\ell}\,, \tag{16}$$

whenever a particular solution for the body force problem, as stated in eqn (1), is available. By the same token, the vector $b_m$ of equivalent nodal displacements, introduced in eqn (13), may be approximated as developed in the following:

$$\begin{gathered}\int_{\Omega} b_i u^{*}_{im} \mathrm{d}\Omega = -\int_{\Gamma} \sigma^{p}_{ji} \eta_j u^{*}_{im} \mathrm{d}\Gamma + \int_{\Gamma} \sigma^{*}_{jim} \eta_j u^{p}_{i} \mathrm{d}\Gamma + \delta_{im} u^{p}_{i} \\ \Rightarrow\quad b_i \approx -G_{m\ell} t^{p}_{\ell} + H_{mn} d^{p}_{n}\,,\end{gathered} \tag{17}$$

since, for a sufficiently refined boundary mesh, the displacements $u^{p}_{i}$ and the traction forces $t^{p}_{i} = \sigma^{p}_{ji}\eta_j$, related to an arbitrary particular solution of the non-homogeneous governing eqn (1), whenever available [7], can be approximated accurately enough by nodal displacement and traction force parameters $d_n^p$ and $t_\ell^p$ in terms of the interpolation functions of eqns (10, 11):

$$\begin{aligned} u_i^p &\approx u_{in} d_n^p, \\ \sigma_{ji}^p \eta_j &\approx t_{i\ell} t_\ell^p \quad \text{on} \quad \Gamma. \end{aligned} \tag{18}$$

Then, making use of eqns (14, 17), a convenient way of expressing eqn (13) is

$$\mathbf{H}\left(\mathbf{d}-\mathbf{d}^p\right)=\left(\mathbf{G}+\mathbf{C}^{\mathrm{T}}\mathbf{R}^{\mathrm{T}}\right)\left(\mathbf{t}-\mathbf{t}^p\right). \tag{19}$$

One identifies in eqn (14), as supported by other linear algebra manipulations [3], that the columns of the matrix $\mathbf{R}$ in eqn (19) span the space of traction forces $\left(\mathbf{t}-\mathbf{t}^p\right)$ that cannot be transformed (are not in equilibrium). Then,

$$\left(\mathbf{G}+\mathbf{C}^{\mathrm{T}}\mathbf{R}^{\mathrm{T}}\right)\mathbf{R}=\mathbf{0} \quad \Rightarrow \quad \mathbf{C}^{\mathrm{T}}=-\mathbf{G}\mathbf{R}\left(\mathbf{R}^{\mathrm{T}}\mathbf{R}\right)^{-1}, \tag{20}$$

which leads to the consistent boundary element equation

$$\mathbf{H}\left(\mathbf{d}-\mathbf{d}^p\right)=\mathbf{G}_a\left(\mathbf{t}-\mathbf{t}^p\right)\equiv\mathbf{G}\mathbf{P}_R^{\perp}\left(\mathbf{t}-\mathbf{t}^p\right), \tag{21}$$

where $\mathbf{G}_a \equiv \mathbf{G}\mathbf{P}_R^{\perp}$ is the admissible part of $\mathbf{G}$ and

$$\mathbf{P}_R^{\perp}=\mathbf{I}-\mathbf{P}_R=\mathbf{I}-\mathbf{R}\left(\mathbf{R}^{\mathrm{T}}\mathbf{R}\right)^{-1}\mathbf{R}^{\mathrm{T}} \tag{22}$$

is the orthogonal projector onto the *admissible* space of the traction forces, which comprises the subsets of traction forces that are in balance and can therefore be transformed into equivalent nodal displacements via the flexibility matrix $\mathbf{G}_a$.

### 2.4 Spectral assessment of the matrices involved in the BEM

Let $\mathbf{W}=\left[W_{ns}\right]\in R^{n^d \times n^r}$ be a matrix whose columns form an orthogonal basis of the nodal displacements $\mathbf{d}$ of eqn (19) related to rigid-body displacements, such that $\mathbf{W}^{\mathrm{T}}\mathbf{W}=\mathbf{I}$. Then, the rigid-body displacement functions $u_{is}^r$ introduced in eqn (7) may be normalized in such a way that their nodal values coincide with $W_{ns}$ at the nodal points, following that

$$u_{is}^r = u_{in} W_{ns} \quad \text{on} \quad \Gamma. \tag{23}$$

Moreover, it is sometimes advisable to think of the boundary traction forces as expressed in terms of equivalent nodal forces $\mathbf{p}=\left[p_n\right]\in R^{n^d}$ that come up from the virtual work statement

$$\delta d_m p_m = \delta d_m \int_\Gamma u_{im} t_{i\ell} \mathrm{d}\Gamma t_\ell$$
$$\Rightarrow \quad p_m = L_{\ell m} t_\ell \quad \text{or} \quad \mathbf{p} = \mathbf{L}^{\mathrm{T}}\mathbf{t}, \tag{24}$$

where $\mathbf{L}^{\mathrm{T}}$, as defined above, performs an equilibrium transformation.

With the definitions of $\mathbf{W}$ and $\mathbf{L}^{\mathrm{T}}$, given above, one checks the equivalence

$$\mathbf{R} \equiv \mathbf{LW}, \tag{25}$$

for $\mathbf{R}$ defined as in eqn (15), which means that, for a finite domain,

$$\mathbf{W}^{\mathrm{T}}(\mathbf{p} - \mathbf{p}^p) = \mathbf{0} \quad \Leftrightarrow \quad \mathbf{R}^{\mathrm{T}}(\mathbf{t} - \mathbf{t}^p) = \mathbf{0}. \tag{26}$$

The relations above help to shed light on the spectral properties of the matrices $\mathbf{H}$ and $\mathbf{G}_a$ of eqn (21). $\mathbf{W} = N(\mathbf{H})$ and $\mathbf{G}_a\mathbf{R} = \mathbf{0}$ is a partial consistency check. Defining $\mathbf{V}$ as the null space $\mathbf{V} = N(\mathbf{H}^{\mathrm{T}})$, one checks that $\left|\mathbf{V}^{\mathrm{T}}\mathbf{G}_a\right| \approx 0$, and not $\left|\mathbf{V}^{\mathrm{T}}\mathbf{G}_a\right| = 0$, which means that eqn (21) is not completely consistent. This is expected, as eqn (21) is obtained from a weighted-residuals statement, eqn (4), that is not variationally consistent – just compare it with eqn (3).

### 2.5 On the numerical solution of the problem

The inconsistent version of eqn (21),

$$\mathbf{H}(\mathbf{d} - \mathbf{d}^p) = \mathbf{G}(\mathbf{t} - \mathbf{t}^p), \tag{27}$$

obtained by neglecting the residual error $\boldsymbol{\varepsilon}$ in eqn (13), may be expressed as

$$\mathbf{Ax} = \mathbf{y}, \tag{28}$$

where the vector $\mathbf{x}$ gathers all the unknown coefficients of $\mathbf{d}$ and $\mathbf{t}$; $\mathbf{y}$ is the vector of known quantities; and the non-symmetric matrix $\mathbf{A}$ is obtained by adequately collecting the columns of $\mathbf{H}$ and $\mathbf{G}$ corresponding to the unknown coefficients [1]. Although $\mathbf{G}$ is a rectangular matrix, $\mathbf{A}$ becomes square if the boundary conditions are adequately introduced [8]. If one is lucky, $\left|\mathbf{V}^{\mathrm{T}}\mathbf{G}\right| >> 0$ and $\mathbf{A}$ is well conditioned. However, there is an in principle uncontrolled amount of rigid-body displacements implicit in the term $u^*_{im}$ of eqn (7), so that the good conditioning of $\mathbf{A}$ cannot be assured beforehand [9].

The solution of eqn (28), if obtained from eqn (21), leads to a matrix $\mathbf{A}$ that is by construction ill conditioned. However, inclusion of the restriction $\mathbf{R}^{\mathrm{T}}(\mathbf{t} - \mathbf{t}^p) = \mathbf{0}$, eqn (26), enables the establishment of an equation system that is always well conditioned, to be solved in terms of generalized inverses. As several numerical tests have shown [3, 8], results using either $\mathbf{G}$ or $\mathbf{G}_a$ are comparable in terms of accuracy, provided that $\mathbf{G}$ does not become ill conditioned. Although the residual $\boldsymbol{\varepsilon}$ of eqn (13) tends to vanish with increasing mesh refinement, an eventual ill-conditioning of $\mathbf{G}$, once present, persists independently from mesh refinement.

### 2.6 Evaluation of a stiffness matrix

Instead of solving a problem in terms of the transformed system of eqn (28), one may need to obtain a formulation in terms of a stiffness-type matrix,

$$\mathbf{K}\left(\mathbf{d}-\mathbf{d}^{p}\right)=\left(\mathbf{p}-\mathbf{p}^{p}\right). \tag{29}$$

The matrix $\mathbf{K}$ is obtained by solving for $\left(\mathbf{t}-\mathbf{t}^{p}\right)$ in either eqn (21) or (27) and making use of $\mathbf{L}^{\mathrm{T}}$ defined in eqn (24). A first possibility comes from eqn (27),

$$\mathbf{K}_{inconsistent\_1}=\mathbf{L}^{\mathrm{T}}\mathbf{L}(\mathbf{GL})^{-1}\mathbf{H}, \tag{30}$$

where $\mathbf{L}(\mathbf{GL})^{-1}$ is a {1, 2, 3}-inverse of $\mathbf{G}$ ($\mathbf{GL}$ supposedly well conditioned). An alternative is to resort to the {1, 2, 3, 4}-inverse $\mathbf{G}^{\mathrm{T}}\left(\mathbf{GG}^{\mathrm{T}}\right)^{-1}$, obtaining

$$\mathbf{K}_{inconsistent\_2}=\mathbf{L}^{\mathrm{T}}\mathbf{G}^{\mathrm{T}}\left(\mathbf{GG}^{\mathrm{T}}\right)^{-1}\mathbf{H}. \tag{31}$$

Either construction above leads to forces that are not in equilibrium. A conceptual improvement is

$$\mathbf{K}_{inconsistent\_3}=\mathbf{L}^{\mathrm{T}}\mathbf{P}_{R}^{\perp}\mathbf{L}(\mathbf{GL})^{-1}\mathbf{H}. \tag{32}$$

However, the only fully consistent formulation stems from eqn (21) [8],

$$\mathbf{K}_{consistent}=\mathbf{L}^{\mathrm{T}}\mathbf{P}_{R}^{\perp}\mathbf{L}\left(\mathbf{G}_{a}\mathbf{L}+\mathbf{WW}^{\mathrm{T}}\right)^{-1}\mathbf{H}. \tag{33}$$

## 3 A subtle simplification and improvement

The functions $u_{in}$ and $t_{i\ell}$ that interpolate displacements and traction forces, as introduced in eqns (10, 11) for curved, isoparametric elements, are usually polynomials of the boundary natural coordinates, a usage possibly borrowed from the finite element tradition. This is fine for the displacements (representation of rigid-body displacements is a requirement) and suited for the evaluation of the matrix $\mathbf{H}$ defined in eqns (12, 13), since the Jacobian $|J|$ of the parametric transformation cancels out in the product $\eta_j \mathrm{d}\Gamma$, whether a 2D or 3D problem is being modeled, and the integrand consists of low-order polynomials that multiply the kernel $\sigma^{*}_{ijm}$.

However, there is no justification for the traction forces to be interpolated by a polynomial along a curved boundary, since, because of the term $\eta_j$ in eqn (2), the boundary forces vary according to the inverse of $|J|$ (there is not such a requirement as the representation of constant-traction forces along the boundary). Then, one suggests the replacement of the polynomials $t_{i\ell}$ in eqn (11) by (a caveat: do not use together with the quarter-point trick in fracture mechanics)

$$t_{i\ell} \leftarrow \frac{|J|_\ell}{|J|} t_{i\ell}, \tag{34}$$

where $|J|_\ell$ is the value of the Jacobian at the point characterized by the subscript $\ell$. Nothing changes formally in the developments outlined for the CBEM, except that the numerical integration of the matrix $\mathbf{G}$ becomes much easier. In fact, $|J|$ cancels out in the product $t_{i\ell}\mathrm{d}\Gamma$, according to eqns (12, 13), for $t_{i\ell}$ defined as above, and the only term to be approximated by a polynomial, in the numerical integration of $\mathbf{G}$, is the kernel $u^*_{im}$. Moreover, $L_{\ell m} = \int_\Gamma u_{im} t_{i\ell} \mathrm{d}\Gamma$ in eqn (24) now involves only low-order polynomials in the integrand.

## 4 Numerical examples

Figure 1 illustrates an irregularly-shaped continuum for a set of five 2D numerical examples of potential problem, which presents the same conceptual issues as the ones of an elasticity problem. The mesh shown corresponds to examples M2_quad and M2_quadN, with a total of 62 quadratic elements and 124 nodes (46 elements along the external boundary and 16 elements modeling the hole indicated). There are also examples M1_quad and M1_quadN, with 23 quadratic elements and 46 nodes only along the external boundary (no hole is modeled), as well as example M1_lin with the same mesh as before, but comprising 46 linear elements. In the examples M2_quad and M1_quad, polynomials $t_{i\ell}$ are used, as initially proposed in eqn (11). In the examples M2_quadN and M1_quadN, one uses $t_{i\ell}$ according to eqn (34).

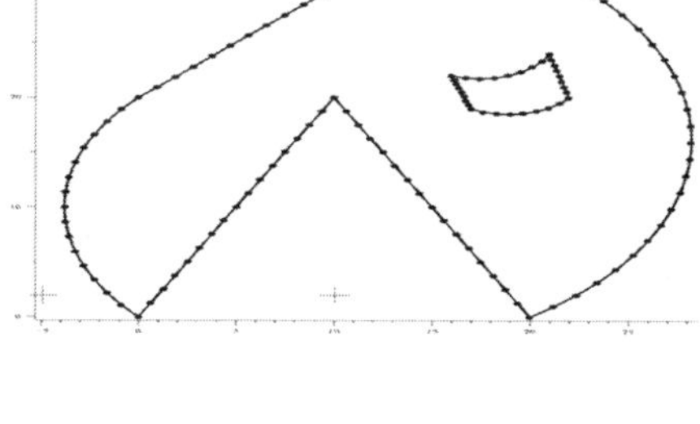

Figure 1: Discretization scheme to illustrate five numerical examples [4].

The first row of Table 1 shows the Euclidean norms of the matrix product $|\mathbf{K}^{\mathrm{T}}\mathbf{W}|$ for the examples under analysis, where $\mathbf{K}$ is the inconsistent matrix defined in eqn (30), to check that the lack of equilibrium of the 'stiffness' system derived from the CBEM is not so relevant for a sufficiently fine mesh, as already known in the literature. (Such an Euclidean norm for the 'stiffness' matrices of eqns (32, 33) is equal to zero by construction.) One also shows, in the second row, the Euclidean norms of the matrix product $|\mathbf{G}_a^{\mathrm{T}}\mathbf{V}|$, for $\mathbf{G}_a$ as in eqn (21), to attest the comment made in the paragraph after eqn (26) as well as to corroborate the outlines of Sections 2.5 and 2.6.

Table 1: Euclidean norms of two matrix products.

| | M1_lin | M1_quad | M1_quadN | M2_quad | M2_quadN |
|---|---|---|---|---|---|
| $\|\mathbf{K}^{\mathrm{T}}\mathbf{W}\|$ | .688e-3 | .167e-2 | .168e-2 | .798e-3 | .799e-3 |
| $\|\mathbf{G}_a^{\mathrm{T}}\mathbf{V}\|$ | .545e-2 | .442e-2 | .448e-2 | .232e-2 | .233e-2 |

One also runs a series of patch tests for a total of ten potential fields applied to the models, eight of them, numbered 1 to 8, are listed in the vector $\langle x \quad y \quad xy \quad x^2-y^2 \quad x^3-3xy^2 \quad -3x^2y+y^3 \quad x^4+y^4-6x^2y^2 \quad x^3y-xy^3 \rangle$, besides potential fields number 9 and 10 that correspond to $\ln(r)/2\pi$, where $r$ is the distance to the source points (-5, 2) and (10, 2) marked as two crosses in Figure 1. The accuracy of the solution with the inconsistent matrix $\mathbf{G}$, as given in eqn (27), is assessed in the first graphic of Figure 2, for all ten potential fields, with corresponding values of $\mathbf{d}$ and $\mathbf{t}$ evaluated analytically. The results using $\mathbf{G}_a$, as in eqn (21), are almost indistinguishable from the ones with $\mathbf{G}$.

The second graphic in Figure 2 assesses the accuracy of using $\mathbf{L}^{\mathrm{T}}$, as given in eqn (24), to evaluate equivalent nodal gradients, for $\mathbf{p}$ directly integrated as $p_m = \int_\Gamma u_{im} t_i \mathrm{d}\Gamma$. The last graphic assesses the accuracy of the 'stiffness' system of eqn (29), where $\mathbf{K}$ is the inconsistent matrix defined in eqn (30). Results using the alternative definitions of $\mathbf{K}$ in eqns (31-33) are almost indistinguishable from these ones. The generalized inverses used to arrive at $\mathbf{K}$ in eqns (30-33) deserve further numerical investigation. In fact, space restrictions of the manuscript prevent the elaboration of many interesting conclusions from these simple examples and graphics.

The curved parts of the boundaries are very deleterious to the numerical accuracy, particularly affecting the error norm $|\mathbf{L}^{\mathrm{T}}\mathbf{t}-\mathbf{p}|$. One checks that the suggestion expressed in eqn (34) significantly improves the problems' response to constant gradients, in terms of both $|\mathbf{L}^{\mathrm{T}}\mathbf{t}-\mathbf{p}|$ and $|\mathbf{Hd}-\mathbf{Gt}|$, but does not lead to perceptible improvements when testing for higher order gradients. However, the simplification achieved with eqn (34) regarding numerical implementation is